Spring Boot 应用开发实战

饶仕琪 著

清华大学出版社
北京

内 容 简 介

随着移动互联网的发展，对Web开发的需求日益上升。Spring Boot作为Web开发领域中的利器，无论是单体应用，还是用于面向服务架构或者微服务架构，都有不错的表现。本书面向准备在Web开发领域一展拳脚的读者，详细介绍了Spring Boot 2.3应用开发的相关知识。

本书共10章，第1~2章介绍了Spring Boot 2.3开发基础，包括环境搭建、依赖引入以及如何从零到一地构建一个基础Spring Boot应用；第3~7章以章为单位聚焦不同技术领域，包括Web服务、数据持久化、服务安全性、测试验证以及部署运维，结合详细示例讲解各领域内通用的解决方案；第8~10章为实战内容，包括三种类型项目的实战：聊天服务、在线商城、个人云盘。

本书内容全面、实例丰富，非常值得广大Web开发从业者作为案头的参考书，也可作为高等院校计算机及相关专业的教材或课程设计参考书。

本书封面贴有清华大学出版社防伪标签，无标签者不得销售。
版权所有，侵权必究。举报：010-62782989，beiqinquan@tup.tsinghua.edu.cn。

图书在版编目（CIP）数据

Spring Boot 应用开发实战 / 饶仕琪著.—北京：清华大学出版社，2021.3(2023.8重印）
ISBN 978-7-302-57526-9

Ⅰ.①S… Ⅱ.①饶… Ⅲ.①JAVA语言－程序设计 Ⅳ.①TP312.8

中国版本图书馆 CIP 数据核字（2021）第 025257 号

责任编辑：夏毓彦
封面设计：王　翔
责任校对：闫秀华
责任印制：丛怀宇

出版发行：清华大学出版社
　　　　　网　　址：http://www.tup.com.cn，http://www.wqbook.com
　　　　　地　　址：北京清华大学学研大厦A座　　邮　编：100084
　　　　　社 总 机：010-83470000　　　　　　　　邮　购：010-62786544
　　　　　投稿与读者服务：010-62776969，c-service@tup.tsinghua.edu.cn
　　　　　质 量 反 馈：010-62772015，zhiliang@tup.tsinghua.edu.cn
印 装 者：大厂回族自治县彩虹印刷有限公司
经　　销：全国新华书店
开　　本：190mm×260mm　　印　张：20.75　　字　数：531千字
版　　次：2021年3月第1版　　　　　　　　　印　次：2023年8月第2次印刷
定　　价：79.00元

产品编号：086671-01

前　　言

现代人越来越离不开互联网。打车、购票软件助力出行，支付软件让各类交易变得快捷，电商与物流系统的联合让北方的居民也能吃上南国的水果。互联网已经涉及现代生活的方方面面，Web 开发技术为之贡献了许多。Spring 作为 Java Web 开发的中坚力量，在该领域的快速发展中扮演着重要的角色。Spring 社区不断地向外输出 Web 开发技术，在保证性能的同时兼顾开发效率。这样的特点让基于 Spring 的 Web 开发技术得到广泛的应用，如果选择使用 Java 语言开发 Web 服务，Spring 就是 Java 开发人员的首选。

Spring Boot 是 Spring 社区中的顶级项目，在整个生态中如同基石一样。无论是想结合模板引擎实现一个单体应用、支撑前端项目的 RESTful 服务或者基于 Spring Cloud 开发一套微服务，这些都离不开 Spring Boot。

本书从实际应用出发，理论结合实例，深入浅出地对 Spring Boot 开发进行讲解。实战内容将贯穿全书，指导读者通过动手实践，从一行语句、一个方法到整个的项目，完整地理解 Spring Boot 开发的流程，从而获得和提升 Web 应用开发的能力。

本书特色

1. 涵盖 Spring 生态中的主流框架

框架的选择需要足够慎重，好的框架不仅能解决开发过程中的问题，还能保证在项目运行的过程中尽可能不制造新的问题。本书中包含 Spring Boot 整合各类框架的内容，这些框架基本来自于 Spring 团队或是由 Spring 团队所推荐的解决方案，足够主流，也足够可靠。

2. 根据程序实现经历的周期展开对 Spring Boot 开发的讲解

全书内容根据一个 Web 项目的实现周期展开，从前期的选型以及项目搭建到具体各种模块的开发，从对程序的测试到最终项目的监控部署，完整地描述项目周期，为广大读者提供更多的视角，以提高本书的参考价值。

3. 讲解过程中穿插实战，覆盖不同读者群体

本书在讲解过程中穿插着丰富的示例以及实战内容。从本书的前半部开始，边讲解边逐步完善一个博客程序；到本书最后的实战阶段，完整地实现三个实战项目，并针对特定解决方案讲解附加的示例代码，相关源码可通过网盘下载。这样的方式既可以让新手读者逐行语句进行实践，也可以让基础扎实的读者了解与参考实现当中的细节。

4. 模块驱动，应用性强

本书当中的解决方案根据需求与场景进行区分，可以根据需求快速阅读并实践，帮助读者实实在在地解决问题。

本书内容体系

第 1 章 Spring Boot 基础知识

本章介绍 Spring Boot 开发有关的基础知识。主要包含工具选择、环境搭建、项目构建以及基础的开发流程。本章的内容主要为了帮助读者掌握 Spring Boot 开发所需的基本概念。

第 2 章 使用 Spring Boot 构建 Web 应用程序

本章介绍构建基于 Spring Boot 的单体应用所需掌握的基本知识。主要包括数据持久化、模板框架、文件上传等内容。

第 3 章 创建 RESTful Web 服务

本章专注于讲解如何构建一个 RESTful Web 服务。讲解过程中将涉及 HTTP 的基础概念以及 RESTful Web 服务的设计思路，帮助读者在理解实现流程之外，还能理解承载这些解决方案的原理。

第 4 章 数据库与持久化技术

本章着重探讨数据持久化技术。依次讲解了简单易懂的 JdbcTemplate、方便快捷的 ORM 解决方案 JPA 以及两种不同的 NoSQL——MongoDB 和 Redis。整个过程层层递进，帮助读者理解不同场景下该使用关系型数据库还是非关系型数据库，以及不同数据库在 Spring Boot 中的调用方法。

第 5 章 应用程序安全性

本章的切入点是应用程序的安全。Spring Boot 的安全可以通过整合 Spring Security 框架来实现。在介绍完 Spring Security 的整合流程之外，还介绍了 Session-Cookie 与 Token 两种典型的鉴权模式。不同的鉴权模式之间并无绝对的优劣之分，与持久化技术一样，没有最好的方案，只有适合的方案。

第 6 章 自动化测试

自动化测试在开发过程中处于常常被人忽略的地位。一方面编写完整有效的自动测试用例与编写出可靠的程序一样，需要耗费很大精力；另一方面，测试的收益并不明显。本章介绍了不同粒度的自动化测试方案，阐述了自动化测试对整个项目的意义，并且讲解了如何通过 Spring Boot 提供的测试框架快速实现测试用例。

第 7 章 运维与部署

Spring Boot 不仅开发起来很高效，部署起来也非常友好。本章介绍 Spring Boot 支持的运维与部署方案以及开发者工具，帮助开发人员获得更好的开发体验。

第 8 章 实战 1：基于 STOMP 协议的聊天服务

本章开始便是本书的实战环节，介绍如何从零到一地开发一个聊天服务，其中包含 STOMP

通信协议的原理以及后端服务的实现。

第9章 实战2：在线商城

本章实战内容主要关于如何实现一个商城服务。这类相对传统的Web服务依赖于页面的表现，在后端数据库以及程序的设计外，还详细介绍了模板引擎Thymeleaf的使用。

第10章 实战3：个人云盘

个人云盘项目用到了成熟的第三方中间件MinIO，通过MinIO的存储功能作为个人云盘的核心功能。为实现可快速重复地集成MinIO，本章还介绍如何针对第三方服务封装Starter，通过引入Starter模块实现对第三方服务的快速集成。

源码下载与技术支持

本书配套的源码，请用微信扫描右边二维码获取，可按页面提示，发到邮箱中下载。阅读过程中如果有疑问，请联系booksaga@163.com，邮件主题为"Spring Boot 应用开发实战"。

适合阅读本书的读者

- 需要全面学习Spring Boot开发技术的人员
- Web开发程序员
- Java程序员
- Java EE开发工程师
- 希望提高项目开发水平的人员
- 专业培训机构的学员
- 软件开发项目经理
- 需要一本案头必备查询手册的人员

作者
2021年1月

目　　录

第 1 章　Spring Boot 基础知识 1

1.1　Spring 与 Spring Boot 1
1.1.1　当我们谈论 Spring 时会谈论些什么 1
1.1.2　什么是 Spring Boot 2
1.1.3　Spring Boot 的优势 3
1.2　Spring Boot 2.3 开发环境 4
1.2.1　选择合适的 JDK 4
1.2.2　选择趁手的 IDE 5
1.2.3　选择适用于大型项目的自动化构建工具 7
1.3　Spring Initializr 初始化项目 8
1.3.1　什么是 Spring Initializr 8
1.3.2　开始吧！start.spring.io 8
1.3.3　使用 IDE 初始化 Spring Boot 工程 10
1.3.4　初探 Spring Boot CLI 12
1.4　Spring Boot 目录结构 13
1.4.1　初始化的工程结构 13
1.4.2　推荐的工程结构 15
1.4.3　Maven Wrapper 让构建工具随源码分发 17
1.5　构建第一个 Spring Boot 项目 18
1.5.1　经典"Hello World" 18
1.5.2　使用 JUnit 5 测试 19
1.5.3　创建工具类 CommonUtil 21
1.5.4　使用 JPA 进行数据持久化 22
1.5.5　修改控制器以及对应模板文件 25
1.5.6　发布 HTTP 接口 29
1.6　Spring Boot 自动配置与外部配置 31
1.6.1　自动配置 31
1.6.2　外部配置 32
1.6.3　命令行配置 32
1.6.4　application.yaml/properties 配置文件 33

第 2 章 使用 Spring Boot 构建 Web 应用程序 36

- 2.1 实体与数据持久化 36
  - 2.1.1 数据持久化框架 36
  - 2.1.2 什么是实体 37
  - 2.1.3 浅谈 Spring Data JPA 38
  - 2.1.4 使用 Lombok 简化 POJO 40
- 2.2 MVC 与模板引擎 42
  - 2.2.1 MVC 架构 42
  - 2.2.2 Mustache 模板引擎 42
  - 2.2.3 构建 MVC 架构的 Web 应用 46
- 2.3 文件上传与下载 50
  - 2.3.1 文件上传 50
  - 2.3.2 文件下载 53
- 2.4 Spring Boot 日志 55
  - 2.4.1 使用预设配置 55
  - 2.4.2 基础配置 56
  - 2.4.3 详细配置 57
  - 2.4.4 Lombok 注解：@Sl4j 和@Commonslog 58
  - 2.4.5 在 Windows 平台输出彩色日志的 JANSI 59
- 2.5 过滤器与拦截器 59
  - 2.5.1 过滤器 60
  - 2.5.2 使用过滤器实现访问控制 60
  - 2.5.3 拦截器 62
  - 2.5.4 使用拦截器记录请求参数 63
- 2.6 Spring Boot 事件 64
  - 2.6.1 事件驱动模型 64
  - 2.6.2 内置事件 65
  - 2.6.3 监听内置事件 65
  - 2.6.4 自定义事件 66
  - 2.6.5 异步事件 68

第 3 章 创建 RESTful Web 服务 69

- 3.1 HTTP 动词 69
  - 3.1.1 构建一个基础的 RESTful Web 服务 70
  - 3.1.2 是 GetMapping 吗？是 RequestMapping 74
- 3.2 请求与响应 76
  - 3.2.1 HTTP 报文 76
  - 3.2.2 简单请求与@RequestParam 77

目录

	3.2.3	使用@PathVariable 获取 URL 中的参数	79
	3.2.4	借助@RequestHeader 读取请求头	80
	3.2.5	@RequestBody 与@ResponseBody	81
	3.2.6	使用 ResponseEntity 处理 HTTP 响应	82
3.3	参数验证		83
	3.3.1	基础验证 Bean Validation	83
	3.3.2	高级验证 Spring Validation	84
	3.3.3	自定义校验	86
3.4	错误处理		88
	3.4.1	使用@ExceptionHandler 处理异常	88
	3.4.2	使用 HandlerExceptionResolver 处理异常	89
	3.4.3	使用@ControllerAdvice 处理异常	90
	3.4.4	抛出 ResponseStatusException 异常	92
3.5	Swagger 文档		92
	3.5.1	Swagger/OpenAPI 规范	92
	3.5.2	生成接口文档	93
	3.5.3	使用注解生成文档内容	94

第 4 章 数据库与持久化技术 97

4.1	使用 JdbcTemplate 访问关系型数据库	97
	4.1.1	引入依赖	97
	4.1.2	准备数据	98
	4.1.3	queryForObject()方法	99
	4.1.4	使用 RowMapper 映射实体	99
	4.1.5	使用 BeanPropertyRowMapper 映射	100
	4.1.6	queryForList()方法	101
	4.1.7	不同的 JDBCTemplate 实现 NamedParameterJdbcTemplate	101
	4.1.8	update()方法	102
4.2	JPA 与关系型数据库		103
	4.2.1	什么是 JPA	103
	4.2.2	再谈 Spring Data JPA	104
	4.2.3	基于 JpaRepository 接口查询	106
	4.2.4	基于 JpaSpecificationExecutor 接口查询	109
	4.2.5	基于 JPQL 或 SQL	113
	4.2.6	多表连接	113
	4.2.7	级联操作	117
	4.2.8	加载类型	117
4.3	Spring Data MongoDB		118
	4.3.1	准备工作	118

  4.3.2 使用 MongoTemplate 访问 MongoDB 120
  4.3.3 使用 MongoRepository 访问 MongoDB 124
4.4 Spring Data Redis 125
  4.4.1 准备工作 125
  4.4.2 使用 RedisRepository 访问 Redis 126
  4.4.3 使用 RedisTemplate 访问 Redis 128

第 5 章　应用程序安全性 130

5.1 基于 Spring Security 的注册登录 130
  5.1.1 Spring Security 简介 130
  5.1.2 用户注册 131
  5.1.3 用户登录 133
  5.1.4 "记住我"功能 135
5.2 权限管理 137
  5.2.1 权限与角色 137
  5.2.2 权限管理体系中的实体：用户、角色与权限 137
  5.2.3 配置与应用 140
  5.2.4 权限管理注解 141
5.3 Session-Cookie 143
  5.3.1 什么是 Session-Cookie 143
  5.3.2 使用 Spring Session 管理 Session 144
  5.3.3 Session 并发配置 146
  5.3.4 强制下线 147
5.4 JWT（JSON Web Token） 148
  5.4.1 关于 JWT 148
  5.4.2 JWT 工作流程 149
  5.4.3 Spring Security 集成 JWT 150
5.5 OAuth 2.0 156
  5.5.1 OAuth 2.0 简介 156
  5.5.2 授权模式 157
  5.5.3 集成 OAuth 2.0 实现 SSO 单点登录 160

第 6 章　自动化测试 164

6.1 单元测试 164
  6.1.1 测试金字塔 164
  6.1.2 JUnit 基础 165
  6.1.3 JUnit 5 简介 170
6.2 断言 172
  6.2.1 assert 关键字 173

6.2.2 JUnit 4 里的断言 173
6.2.3 assertThat 方法 174
6.2.4 自定义 Hamcrest 匹配器 175
6.2.5 断言框架 AssertJ 177
6.3 测试中的模拟行为 Mock 179
6.3.1 测试替身 179
6.3.2 Mockito 框架 180
6.4 集成测试 183
6.4.1 @WebMvcTest 注解 183
6.4.2 @DataJpaTest 注解 187
6.4.3 @SpringBootTest 以及其他一些注解 189

第 7 章 部署与运维 190

7.1 发布与部署 190
7.1.1 Spring Boot 自身的打包方式 1——可执行 jar 文件 190
7.1.2 Spring Boot 自身的打包方式 2——部署于传统 Web 容器的 war 格式 191
7.1.3 更现代的发布流程 1——基于 Docker 的发布与部署 193
7.1.4 更现代的发布流程 2——基于 RPM 的发布与部署 195
7.1.5 多环境配置 197
7.2 运行监控 198
7.2.1 使用 Spring Boot Actuator 查看运行指标 199
7.2.2 集成 Prometheus 200
7.2.3 Grafana 实现可视化监控 203
7.3 Spring Boot 开发者工具 206
7.3.1 整合 spring-boot-devtools 206
7.3.2 自动配置 207
7.3.3 热部署 207
7.3.4 LiveReload 插件支持静态资源的及时更新 208
7.3.5 全局配置 208
7.3.6 远程应用 209

第 8 章 实战 1：基于 STOMP 协议的聊天服务 211

8.1 架构设计 211
8.2 框架搭建 212
8.3 功能实现 214
8.3.1 了解 WebSocket 协议 214
8.3.2 HTTP 请求升级至 WebSocket 的过程 215
8.3.3 了解 WebSocket 应用场景 216
8.3.4 集成 WebSocket 216

         8.3.5　使用 STOMP 协议实现消息模块 218
         8.3.6　模块配置 225
         8.3.7　注册登录 227
         8.3.8　聊天记录 230
         8.3.9　私聊功能 233
　　8.4　测试与验证 238
         8.4.1　集成测试 238
         8.4.2　手工测试 240

第 9 章　实战 2：在线商城 245

　　9.1　架构设计 245
　　9.2　框架搭建 246
　　9.3　数据库设计 247
　　9.4　功能实现 248
         9.4.1　模板引擎 Thymeleaf 249
         9.4.2　实体类 253
         9.4.3　用户注册 256
         9.4.4　用户登录 260
         9.4.5　主页以及商品列表 263
         9.4.6　购物车 266
         9.4.7　页眉、导航条以及页脚 271
　　9.5　测试与验证 272
         9.5.1　测试数据 273
         9.5.2　集成测试 274
         9.5.3　手工测试 276

第 10 章　实战 3：个人云盘 281

　　10.1　架构设计 281
　　10.2　框架搭建 282
         10.2.1　MinIO 与对象存储 282
         10.2.2　MinIO 部署与使用 283
         10.2.3　项目依赖项与软件包结构 283
　　10.3　数据库设计 285
　　10.4　功能实现 286
         10.4.1　MinIO Java SDK 简介 286
         10.4.2　实现 MinIO Starter 290
         10.4.3　实体类 294
         10.4.4　用户注册 296
         10.4.5　用户登录 298

10.4.6　云盘主页 300
　　　10.4.7　页面配置 306
10.5　测试与验证 308
　　　10.5.1　集成测试 308
　　　10.5.2　手工测试 309

第 1 章

Spring Boot 基础知识

距离 Java 的第一个版本已有 25 个年头，这门语言的生态在这漫长岁月里变得愈发丰富多样。这个 Java 生态中有一个不容忽视的名字，那便是 Spring。Spring 技术极大地提高了开发人员的开发效率，将人们从刀耕火种一下带到了工业时代。本书将介绍 Spring 这个开发利器中最锐利的锋芒——Spring Boot。

本章主要涉及的知识点有：

- Spring 与 Spring Boot 的基本概念
- Spring Boot 开发环境的搭建
- 构建 Spring Boot 项目的基本步骤

1.1 Spring 与 Spring Boot

当学习一门技术时，先对其有一个大概的认识是非常必要的，这样对学习方向的把控很有帮助。本节先来了解一下 Spring 与 Spring Boot 的基本概念，看看它们在开发过程将扮演什么角色，发挥什么作用。

1.1.1 当我们谈论 Spring 时会谈论些什么

在不同的语境中 Spring 蕴含不同的含义。狭义的解释为 Spring 指 Spring Framework，因为这是生态的核心，Spring 起源于此。但随着时间推移，社区基于 Spring Framework 构建了更多其他的项目，这样一来，当人们说到"Spring"时，往往指的是整个 Spring 生态。

Spring 的架构如图 1.1 所示，其核心 Core Container 是一个 IoC（Inversion of Control）容器。

IoC 即控制反转，是一种面向对象的思想，作用在于将对象之间的依赖关系交由框架进行统一管理。具体的实现方式是 DI（Dependency Injection，依赖注入）。简单来说，就是开发人员通过 XML 配置或 JavaConfig 的方式将依赖关系告知容器。容器在"恰当"的时机去创建对象，而不需要开发人员过多的关注。

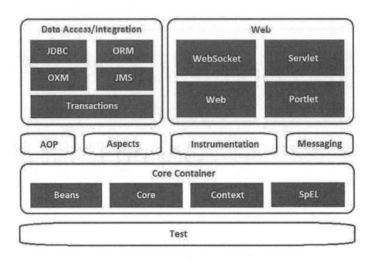

图 1.1　Spring 架构图

Web 模块，指 Web 应用基础功能的集合。其中包含对文件上传的支持、使用 Servlet 监听器初始化 IoC 容器、Web 应用上下文等内容。另外还有对基于 Servlet 开发的支持，这块在 Spring 的体系中又被称作 Spring MVC。在第 2 章将着手构建一个 Web 应用，就离不开 Spring MVC 的支持。

Data Access/Integration，即数据访问与集成方案。JDBC、ORM、OXM 等对于数据库操作的方案被包含其中。在这些模块当中，ORM 将会在之后的章节着重介绍。相较于 JDBC 这样基础的数据库访问方案，使用 ORM 开发起来更为高效。ORM 是对 JDBC 的封装，将字段高效地与对象进行映射，将对数据库的操作转换为对对象的操作。我们将在第 4 章开始学习如何利用这些工具访问数据库。

AOP（Aspect-Oriented Programming，面向切面编程）是通过预编译方式和运行期间动态代理实现程序功能统一维护的一种技术，是 OOP 的延续，也是 Spring Framework 中的一个重要内容，是函数式编程的一种衍生范型。利用 AOP 可以对业务逻辑的各个部分进行隔离，从而使得业务逻辑各部分之间的耦合度降低，提高程序的可重用性，同时提高了开发的效率。

Test 模块提供了 Spring 应用使用 JUnit 和 TestNG 进行单元测试和集成测试的支持。在测试过程中能轻松读取到应用上下文，并且它具有可用于隔离测试代码的 Mock 对象。

1.1.2　什么是 Spring Boot

Spring Boot 是在 Spring 的基础上构建起来的一个项目。它基于"约定优于配置"（Convention Over Configuration）的理念，解决了基于 Spring 开发需要繁复配置的痛点。使用 Spring Boot 进行开发可以巧妙地选择项目所需的依赖项，对依赖中涉及的功能进行自动配置，并且能在不依赖 Web 容器的情况下一键启动，大大简化了应用的开发和部署过程。

以下是 Spring Boot 提供的高级功能：

- 自动配置：根据"starter"依赖项进行自动配置。
- 独立：无需将程序部署到另外的 Web 容器，可通过 run 命令直接启动。
- 智能：配置中的默认值会根据依赖项自动调整。

使用 Spring Boot 可以轻松构建一个企业级的应用并且快速上线，而不用担心配置的准确性和安全性。图 1.2 所示是 Spring Boot 与 Spring Cloud、Spring Cloud Data Flow 的关系。

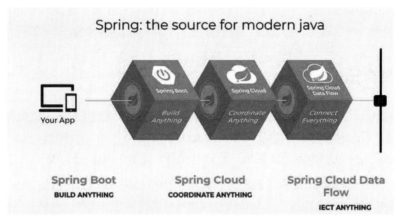

图 1.2　Spring Boot 与 Spring Cloud、Spring Cloud Data Flow 的关系

1.1.3　Spring Boot 的优势

为什么选择 Spring Boot 而不是其他的解决方案？理由有以下几点：

（1）成熟：Spring Boot 基于 Spring Framework。Spring Framework 已经开发超过 15 年，是 J2EE 的轻量级替代方案。

（2）稳定：Spring 生态中的核心模块长期稳定运行，并且它们的更改都向后兼容。开发人员在做版本升级的过程中，不会感到"举步维艰"。

（3）基于 JVM（Java 虚拟机）：Spring 是基于 Java 的，自然依赖于 JVM。JVM 上除了 Java 之外还可以运行其他的语言，例如：Kotlin、Groovy、Scala 等，Spring Boot 同样可以使用这些语言进行开发。

（4）由公司运作的开源项目：这意味着项目可以有规律地更新以及维护有基本的保障。

（5）云原生：Spring Boot 遵循云应用程序的部署原则，并为开箱即用的云做好了准备。它与 Spring Cloud 一起，可以轻松构建分布式系统。

（6）丰富的支持：使用 Spring 可以轻松地将应用连接到不同的关系型数据库、NoSQL、消息队列等中间件。

（7）灵活性：使用 Spring Boot 既可以开发经典的服务端（或称为服务器端，本书统一简称为服务端，以便具有更广义的含义）渲染 Web 应用，也可以开发 RESTful 或者其他形式的 Web-API，甚至可以创建批处理和常规命令行应用程序。

1.2 Spring Boot 2.3 开发环境

在正式编码之前，还需要做一些准备工作。首要任务是挑选并搭建好 Spring Boot 开发环境。一个基础的 Spring Boot 开发配置包括：JDK（Java Development Kit，Java 开发工具集）、IDE（Integrated Development Environment，集成开发环境）以及一款自动化构建工具。得益于开源社区的繁荣，这些配置有不少备选项可供选择。挑选合适的配置是各类开发中绕不开的一个话题。

1.2.1 选择合适的 JDK

目前 Spring Boot 2.3 已对当前最新的 JDK 14 提供了支持，可以在 Spring Boot 中体验到强大又"炫酷"的功能。不过笔者仍推荐使用 JDK 8 或 JDK 11 这两个版本，大多 Java 类库基于这些版本构建，在学习与开发中能够少走很多弯路。除自身的版本之外，JDK 还有发行版之分。

- Oracle JDK。Oracle JDK 称得上是一个经典的选择，Java 归 Oracle 所有，Oracle JDK 自然在市场占有率上占主导地位。在 Oracle 官网即可下载到不同版本的、面向不同操作平台的 JDK 安装程序。
- Liberica JDK。对一般用户而言，Liberica 算得上是最友好的 OpenJDK 发行版了。以 Windows 平台的 JDK 安装为例，许多其他发行版的 JDK 安装过程中都免不了手动配置环境变量，而 Liberica JDK 的 Windows MSI 安装包会自动配置环境变量，并且自动关联 jar 打开方式，省时省力。
- Adopt OpenJDK。一个完全免费的 OpenJDK 版本。这个版本完全免费并且对于 JDK 8 和 JDK 11 提供不超过 4 年的支持。

笔者以 Windows 环境下安装 Liberica JDK 8u252+9 为例，介绍 JDK 的安装步骤：

（1）打开页面 https://bell-sw.com/pages/downloads/#/java-8-lts 下载，如图 1.3 所示。

图 1.3　Liberica JDK 下载页面

（2）单击图中左下角的"Windows"按钮，开始 Windows 平台下的 Liberica JDK 安装包下载。

（3）MSI 格式的安装包下载完成之后，并运行安装包。其中没有需要注意的配置项，一路选择默认项即可。

（4）运行命令行程序 cmd.exe，执行命令"java -version"，如果出现版本信息，即说明 JDK 安装成功，如图 1.4 所示。

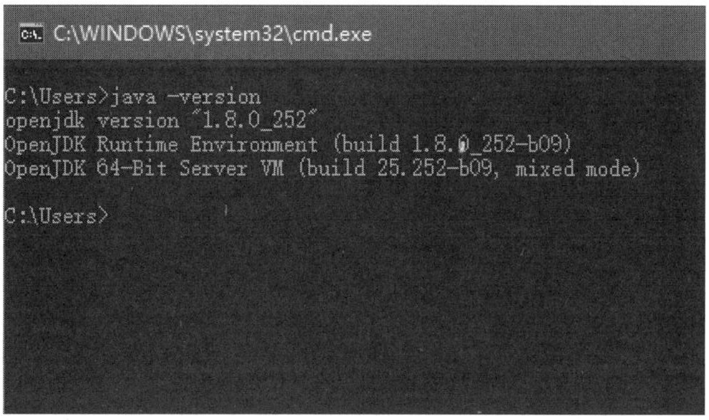

图 1.4　验证 JDK 是否安装成功

1.2.2　选择趁手的 IDE

如果把开发人员比作士兵，那 IDE 就是士兵手中的武器，特别是 Spring Boot 开发过程中对 IDE 的依赖尤为显著。IDE 的选择也是十分丰富，下面列举几款主流的 IDE 供大家选择。

- IntelliJ IDEA（以下简称：IDEA）。目前最流行的 Java IDE 之一，提供诸多功能以提升开发人员的开发体验。笔者在本书介绍的项目构建都是借助 IDEA 来实现的。IDEA 的索引系统是 IDEA 的特色之一，该系统提供更智能的提示以及更便捷的操作。这款 IDE 的优点很多，还需要读者慢慢探索。
- Eclipse。曾经是市场占有率最高的 IDE 之一，具有丰富的插件支持，同样是一款功能强大的 IDE。Spring 社区还在 Eclipse 的基础上提供了 STS 版本（Spring Tool Suite），与 IDEA 相比不遑多让。
- Visual Studio Code（以下简称 VSCode）。严格来说，这款工具虽然称不上是 IDE，但丰富的插件让它与以上两款 IDE 相比也是毫不逊色。基础的插件选择 Java Extension Pack 以及 VS Code 版本的 Spring Tools4 即可开始开发 Spring Boot 应用。

笔者同样以 Windows 平台安装 IDEA 2020.1.3 Community 版本为例，介绍 IDE 的安装及配置流程：

（1）打开官网下载页面：https://www.jetbrains.com/idea/download/#section=windows，如图 1.5 所示。

6 | Spring Boot 应用开发实战

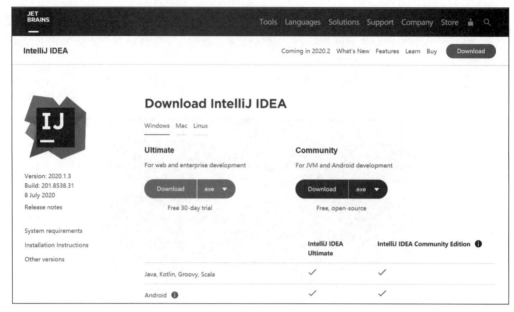

图 1.5　IDEA 下载页面

（2）单击 Community 下的 Download 按钮下载。

（3）下载完成后，运行安装程序。安装过程同样一直选择默认项即可。

（4）打开 IDEA 后即可创建 Java 项目。在窗口的 Project SDK 栏目，可以选择 JDK 版本（见图 1.6）。根据需要可以选择 IDE 自带的 JDK 或者本机安装好的 JDK。

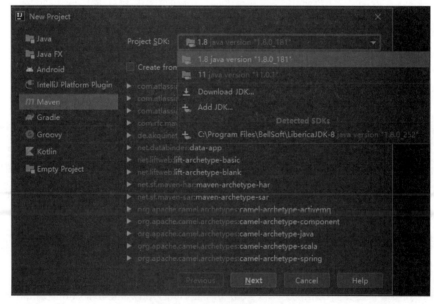

图 1.6　IDEA 创建新项目

1.2.3 选择适用于大型项目的自动化构建工具

当 IDE 准备好之后，理论上来说已经具备了开始编码工作的条件。但实际上，这个过程中经常会遇见很多与编程无关的项目管理工作。比如依赖管理、编译源码、单元测试、项目部署等。当项目的规模并不大的时候，这些工作可以手工实现。等到着手进行大型项目的开发时，特别是使用 Spring Boot 进行企业级的 Web 应用开发，这些工作可能会比编码工作本身更为棘手。这种情况下就需要用到自动化的构建工具来协助我们从前到后地完成编译、连接、打包、依赖管理等工作，一步一步地将源代码打包成可执行的形式。

Java 开发中常见的构建工具有三款：Ant、Maven 和 Gradle。相较于 Maven 与 Gradle，Ant 的历史更为久远。相应地，Ant 不及后来者 Maven 与 Gradle（智能），体验上较为烦琐，渐渐不再流行。因此笔者在这里只着重介绍 Maven 与 Gradle。

Maven 基于 Ant 的项目构建功能解决了构建过程中的两个问题：构建方式及其依赖项。以 Ant 为代表的早期构建工具，在构建过程中需要显式声明各类信息，比如源码路径、输出路径、编译具体步骤等。Maven 在这方面同样使用到"约定优于配置"的理念，对构建方式进行约定，这样一来，配置中需要关注的方面变少许多，相应地提升了开发人员的开发效率。在便利的配置之外，Maven 提出了仓库的概念，可以自动化地管理依赖类库。为了方便 Maven 进行依赖管理以及开发人员对依赖进行配置，Maven 规定在 pom.xml（Maven 配置文件）中需要声明依赖的"坐标"。

```xml
<project>
  <!--Maven 2.x POM 的模型版本始终为 4.0.0-->
  <modelVersion> 4.0.0 </modelVersion>
  <!--项目坐标，即一组唯一标识此项目的值-->
  <groupId> com.mycompany.app </groupId>
  <artifactId> my-app </artifactId>
  <version> 1.0 </version>
  <!--库依赖-->
  <dependencies>
    <dependency>
      <!--所需库的坐标-->
      <groupId> junit </groupId>
      <artifactId> junit </artifactId>
      <version>3.8.1</version>
      <!--此依赖项仅用于运行和编译测试-->
      <scope>测试</scope>
    </dependency>
  </dependencies>
</project>
```

以上是一个 pom.xml 的代码片段，其中由<dependency>包裹的部分便是代表 junit 依赖的坐标。groupId 代表负责维护类库的组织名，artifactId 代表类库对应的项目名，version 则是类库的版本号。这三项合一构成一个完整的 Maven 依赖坐标。

相较于 Maven，Gradle 就更为"新潮"了。它是一款基于多语言开发的自动化构建工具，基于 Ant 和 Maven 的概念引入了基于 Groovy 语言（一种基于 JVM 的敏捷开发语言）的 DSL（Domain-Specific Language，领域特定性语言），而并非 Maven 和 Ant 一直坚守的 XML 形式。采用 DSL 而非 XML 的形式大大提升了配置文件的可阅读性。例如上文的 junit 依赖，在 Gradle 中只需要用一行来表示：

```
compile(junit: junit:3.8.1)
```

除此之外，Gradle 还将构建工具的灵活性提升到了一个新高度，使用 Groovy 语言可以轻松地在 Gradle 配置文件 build.gradle 中修改项目构建的生命周期，类似的操作在 Maven 中需要花费不少时间才能得以实现。

Gradle 虽然如此强大，但并不能说明使用 Gradle 是绝对优于 Maven 的选择。其中缘由十分复杂，还需要在实际开发使用中细细体会。另外，这两款构建工具已内置于 IDEA 中，并且这两款工具可以使用"wrapper"模式，无需另外下载安装。具体的操作内容将在本书第 2 章做详细介绍。

1.3　Spring Initializr 初始化项目

在 Spring 官方的开发指南当中有提到，所有 Spring 应用程序都应该从 Spring Initializr 开始。

1.3.1　什么是 Spring Initializr

Spring Initializr 是 Spring 官方提供的一个项目初始化工具。它是一个可扩展的 API，用以生成基于 JVM 的 Spring Boot 项目的"骨架"，并检查用于生成项目的元数据，例如列出可用的依赖项以及版本号。这个工具有不同的形式，可以单独使用，也可以嵌入其他工具（1.2.2 小节中提到的 IDE 中均有 Spring Initializr 的支持），有 Web UI 的应用，也有命令行的程序。

1.3.2　开始吧！start.spring.io

如果读者仍然不大理解什么是 Spring Initializr，那么可以在浏览器中打开链接：https://st-art.spring.io/来一探究竟。图 1.7 所示是 start.spring.io 对应的 Web 页面。

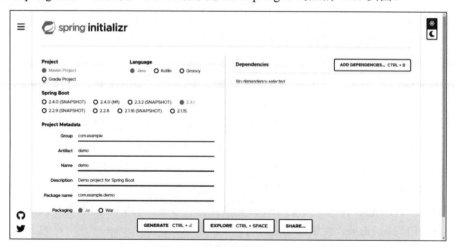

图 1.7　start.spring.io 页面

在页面的 Project 栏中，可以看到了两个熟悉的名词"Maven"以及"Gradle"。用意也十分明显了，在这里可以选择用于构建项目的构建工具。在 Language 栏中，可以选择项目将使用的编程语言。得益于 JVM 的支持，除了经典的 Java 语言，在开发 Spring Boot 时还可以使用 Kotlin 和 Groovy，它们都是非常优秀的 JVM 语言。

在 Spring Boot 这一栏，提供了 Spring Boot 版本的选项。简单介绍一下 Spring Boot 的版本号：

- Release：最终版本。Release 不会以单词形式出现在软件封面上，取而代之的是符号(R)。
- GA：General Availability。正式版本，官方推荐使用此版本，在国外都是用 GA 来说明 Release 版本的。
- RC：Release Candidate，发行候选。该版本已经相当成熟了，基本上不存在导致错误的 BUG，与即将发行的正式版相差无几。
- M：Milestone，又叫里程碑版本。表示该版本较之前版本有功能上的重大更新。
- SNAPSHOT：快照版，可以稳定使用，且仍在继续改进版本。
- Beta：该版本相对于 α 版已有了很大的改进，消除了严重的错误，但还是存在着一些缺陷，需要经过多次测试来进一步消除错误。
- Alpha（不建议使用）：主要是以实现软件功能为主，通常只在软件开发者内部交流，BUG 较多，需要继续修改。
- PRE（不建议使用）：预览版，内部测试版。主要是给开发人员和测试人员测试和查找 BUG 用的。

在图 1.8 中可以看到 Spring Boot 部分版本以及对应标识。在实际学习与开发过程中，笔者更推荐使用相对稳定的 Release 版作为候选项。

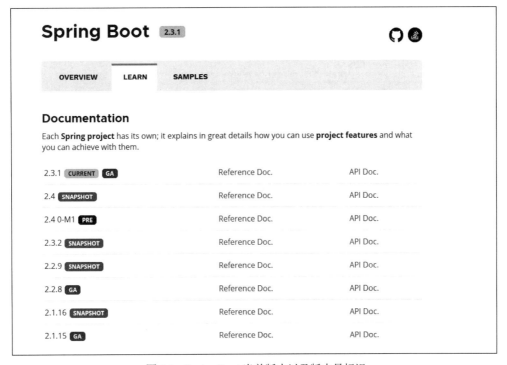

图 1.8　Spring Boot 当前版本以及版本号标识

回到 Spring Initializr 的页面。在 Project Metadata 栏目需要填入将要构建项目的元数据。由上至下依次是：

- Group：项目组织名。
- Artifact：用于构成依赖坐标的项目名。
- Name：用于描述项目的项目名，可以与 Artifact 项保持一致。
- Description：对项目的基本描述内容。
- Package name：包名，通常可以由 Group 与 Artifact 共同组成。
- Packaging：打包方式。
- Java：Java 版本号。

在 Project Metadata 栏中填入以上信息后，Spring Initializr 就能帮助开发人员生成一个基于 Maven 或 Gradle 构建的基本工程了。不过项目难免会需要额外的依赖，那么还需要把目光再投向 Dependencies 这个栏。单击"ADD DEPENDENCIES"按钮，挑选项目所需要的依赖以及开发工具。图 1.9 所示为单击按钮后呈现的页面。

图 1.9　ADD DEPENDENCIES 页面

比如需要 Spring Web 的依赖。在输入框中输入 Web，就可以看到 Spring Web 的候选项。单击 Spring Web 后，这个依赖就被加入到了依赖列表当中。

配置好依赖之后，单击页面左下角的"GENERATE"按钮，一个初始化过后的 Spring Boot 工程就开始下载了。

1.3.3　使用 IDE 初始化 Spring Boot 工程

start.spring.io 固然方便，但并不是人人都乐意于在创建一个新 Spring Boot 项目时都打开浏览器操作一番。不必苦恼，针对 Spring Boot 工程的初始化，各大 IDE 也都有插件提供相关的支持。

下文以 IDEA2020.1.3 Community 版本为例进行讲解，在该版本的 IDEA 中会使用插件"Spring Assistant"来支持 Spring Boot 工程初始化。这个插件需要另行安装，操作步骤如下：

（1）在工具栏中依次选择"File"→"Settings"，打开 IDEA 的设置页面。
（2）在设置页面的左侧选择"Plugins"，打开 IDEA 的插件管理页面。
（3）在搜索栏中输入"Spring Assistant"，即可看到所需的插件。
（4）单击右上角的"Install"按钮安装插件。
（5）待安装完成后，单击"Restart IDE"按钮以重启 IDE。

这样就完成了一个插件的安装。Spring Assistant 的安装页面如图 1.10 所示。

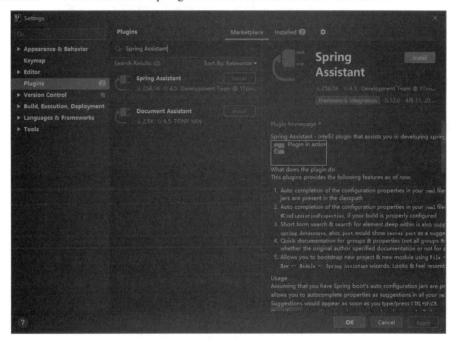

图 1.10 Spring Assistant

在插件安装完成之后，就可以在 IDEA 中初始化一个 Spring Boot 工程了。操作步骤如下：

（1）在工具栏中依次选择"File"→"New"→"Project..."，打开新建项目页面，如图 1.11 所示。

（2）在新建项目页面左侧可以选择"Spring Assistant"选项。对应页面可以配置 Spring Initializr 的服务地址。目前仅需要使用默认的 start.spring.io 即可。单击"Next"按钮进入下一步。

（3）这一步需要配置的内容和在 1.3.2 小节中配置的相同，这里就不再展开了。配置页内容如图 1.12 所示。

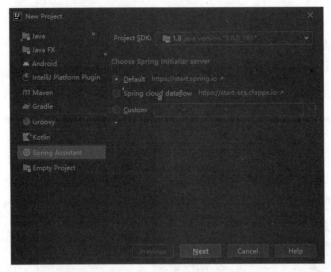

图 1.11　创建新项目页面

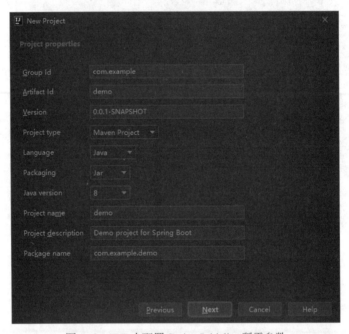

图 1.12　IDE 中配置 Spring Initializr 所需参数

1.3.4　初探 Spring Boot CLI

上文已经介绍了两种方法来初始化 Spring Boot 工程。不过都需要打开页面选择，有没有更高效更"极客"的方法？还真有！Spring 社区为了方便 Spring Boot 开发，推出了一款命令行工具——Spring Boot CLI。当然并不一定需要 CLI 就能着手开发功能强大的 Spring Boot 程序，但这绝对是最快的方法了。

首先，需要安装 Spring Boot CLI，操作步骤如下：

（1）在 Spring Software Repository 里下载 Spring Boot CLI 的发行版。链接地址为 https://repo.spring.io/release/org/springframework/boot/spring-boot-cli/2.3.1.RELEASE/spring-boot-cli-2.3.1.RELEASE-bin.zip。

（2）下载后解压安装包，并把./ spring-2.3.1.RELEASE/bin 路径配置到 Path 环境变量中。

（3）测试是否配置成功。打开 cmd，输入命令"spring --version"。当页面出现 Spring Boot CLI 的版本号时，如"Spring CLI v2.3.1.RELEASE"。这就说明 Spring Boot CLI 已经安装完成了。

使用 Spring Boot CLI 来初始化 Spring Boot 工程，所依赖的是其中的"init"命令，使用 init –list 即可查看可用参数：

```
spring init -list
```

假设要构建一个 Web 应用，其中使用 JPA 实现数据持久化，使用 Spring Security 进行安全管理，可用--dependencies 或者-d 来指定初始依赖：

```
spring init -dweb,jpa,security
```

或者需要用 Gradle 来构建项目。--build 参数将 Gradle 指定为构建工具：

```
spring init -dweb,jpa,security --build gradle
```

无论是 Maven 还是 Gradle 的构建，都会产生一个可执行 jar 文件。但如果需要的是一个 war 格式的文件该怎么办呢？可以通过--packaging 或者-p 参数来解决：

```
spring init -dweb,jpa,security --build gradle -p war
```

细心的读者可能会发现，包括之前用页面初始化项目时，最终的初始化的结果都是一个名为 demo.zip 的压缩包。如果实在不想在之后执行解压操作，Spring Boot CLI 可以帮我们做到这一点吗？答案是肯定的。在命令的最后指定一个目录，文件下载完成后会自动解压这个目录中：

```
spring init -dweb,jpa,security --build gradle -p war myapp
```

至此，一个初始化的 Spring Boot 程序就创建好了。

1.4　Spring Boot 目录结构

根据上一节的介绍，我们已经在本地完成了一个基础 Spring Boot 工程的初始化。一个基础的工程结构包含一个主应用类、一个配置文件以及若干个与构建工具相关的文件。下面将介绍 Spring Boot 工程的目录结构。

1.4.1　初始化的工程结构

首先通过命令"spring init -dweb,jpa,security --build maven -p jar basic-project"创建好一个基础工程。工程的目录结构如图 1.13 所示。

图1.13　Spring Boot 工程初始的目录结构

由上至下的文件以及目录：

（1）.idea：由 IDEA 而非 Spring Initializr 创建。包含该项目的历史记录、版本控制信息等。

（2）.mvn：Maven 的 Wrapper 目录，对应该项目选用的构建工具。如若选用了 Gradle 为项目构建工具，该目录将会被替代为"gradle"。

（3）src：源码目录。包含源码、测试代码以及资源文件目录。

（4）main：主体目录。在主体程序目录下，需要放置主体程序构建所需的代码文件以及资源文件。

（5）test：测试目录。测试目录存放用于构建测试用例相关的代码文件以及资源文件。

（6）java：java 代码目录。在 java 目录之下存放.java 后缀的代码文件。

（7）resources：资源文件目录。在资源文件目录中，存放除代码文件之外任何构建程序所需的"资源文件"，包括但不限于配置文件、模板文件、图片文件等。

（8）DemoApplication：主体程序入口类。

```java
package com.example.basic;

import org.springframework.boot.SpringApplication;
import org.springframework.boot.autoconfigure.SpringBootApplication;

@SpringBootApplication
public class DemoApplication {

 public static void main(String[] args) {
  SpringApplication.run(DemoApplication.class, args);
 }

}
```

在这个类中可以看到熟悉的 main 函数，这说明该类是整个 Spring Boot 应用的入口类。

（9）static 目录与 template 目录：由于初始化的过程中引用了"web"模块，Spring Initializr 创建了这两个目录，以便于构建 MVC（Model-View-Controller，一种软件设计模式）模式的应用。static 主要用于存放 js/css 或者图片之类的静态文件，template 用于存放页面模板文件。

（10）application.properties：Spring Boot 项目的配置文件。还记得前文提到的"约定优先于配置"吗？Spring Boot 提供了许多默认的配置。当默认配置不满足需求时，在配置文件中填写需要的配置项是在 Spring Boot 项目中用于覆盖默认配置的方法之一。

（11）DemoApplicationTests：Spring Boot 项目的测试类。

（12）.gitignore：使用 git 做版本管理的话，需要在该文件内列出忽略文件的列表。

（13）Basic-project.iml：iml 后缀的文件为 IDEA 用于存储开发环境相关信息的配置文件。

（14）HELP.md：md 格式的帮助文档。该文件根据应用程序中所选依赖而定制，包含与这些依赖有关的指南以及参考文档，以便开发人员更好地进行开发。

（15）mvnw 与 mvnw.cmd/gradlew 与 gradle.bat：mvnw 与 gradlew 分别为 Maven Wrapper 以及 Gradle Wrapper 附带的脚本，与（2）中提到的目录文结合起来，以方便用户无须手动安装这些构建工具就能享用它们带来的帮助。

（16）pom.xml/build.gradle：pom.xml 与 build.gradle 分别对应 Maven 和 Gradle 的构建配置文件。需要在其中声明构架项目所需的依赖以及其他配置项。

（17）settings.gradle：如果使用 Gradle 构建项目的话，还会看到 settings.gradle 这样一个文件。作用在于定义所有包含的了模块并标记模块树的目录根。

1.4.2　推荐的工程结构

上一小节介绍了一个初始化的 Spring Boot 目录的结构以及所包含的文件。本小节在此基础上进行延伸，提出一些关于 Spring Boot 工程结构的建议。

> **提　示**
>
> Spring Boot 本身并不会对目录格式进行约束，结构完全可以符合个性并且天马行空。但是一个工程化的项目有越多约定俗成的元素，就更容易上手开发以及学习，这是笔者鼓励初学者根据"最佳实践"进行代码编写的理由。

以下是一个比较经典的 Spring Boot 项目包的结构：

```
com
 +- example
  +- myproject
   +- Application.java
   |
   +- domain
   |  +- Customer.java
   |  +- CustomerRepository.java
   |
   +- service
   |  +- CustomerService.java
```

```
    |
    +- web
    |  +- CustomerController.java
    |
```

- root package：com.example.myproject，所有的类以及其他的 package 都在 root 下。
- 应用主类：Application.java，该类大多直接位于 root package 下。通常会在应用主类加入一些注解以实现一些配置的效果。例如 1.4.2 小节（8）列出的主类中可以看到 @SpringBootApplication 这个注解。该注解为一个复合注解，作用是用于告诉 Spring Boot 该应用已经开启了自动配置以及组件扫描。
- com.example.myproject.domain 包：用于定义实体映射关系与数据访问相关的接口和实现。
- com.example.myproject.service 包：用于编写业务逻辑相关的接口与实现。
- com.example.myproject.web：用于编写 Web 层相关的实现，比如：Spring MVC 的 Controller。

以上便是一个 Spring Boot 工程的推荐结构。root package 与应用主类的位置是整个结构的关键。由于应用主类在 root package 中，因此按照上面的规则定义的所有其他类都处于 root package 下的其他子包中。应用主类中由 @SpringBootApplication 注解开启的自动配置以及扫描，是针对 root package 以及其子包的。如若情况发生变化，例如：

```
com
 +- example
   +- myproject
     +- Application.java
     |
     +- domain
     |  +- Customer.java
     |  +- CustomerRepository.java
     |
     +- service
     |  +- CustomerService.java
     |
   +- web
     |  +- CustomerController.java
     |
```

将 web 目录放置于与 root package 同级。这种情况下 CustomerController.java 无法自动被扫描到，需要使用其他注解对 Spring Boot 工程进行配置。具体的方式有如下两种：

（1）使用 @ComponentScan 注解指定具体的加载包，比如：

```
@SpringBootApplication
@ComponentScan(basePackages="com.example")
public class Application {

    public static void main(String[] args) {
        SpringApplication.run(Bootstrap.class, args);
    }

}
```

(2)使用@Bean 注解来初始化，比如：

```
@SpringBootApplication
public class Application {

    public static void main(String[] args) {
        SpringApplication.run(Bootstrap.class, args);
    }

    @Bean
    public CustomerController customerController() {
        return new CustomerController();
    }

}
```

这些方式可以帮助实现目录结构的自定义，不过为了开发的便利性和可维护性，推荐参考"最佳实践"来组织代码。

1.4.3　Maven Wrapper 让构建工具随源码分发

Spring Initializr 为开发人员提供了以 Wrapper 模式来使用构建工具，其中包含 Maven Wrapper 以及 Gradle Wrapper。这种方式为开发过程带来了极大的便利，因为进一步提高了对开发环境依赖的配置功能。

Maven Wrapper 实质上是依赖一款插件实现的，插件为 takari-maven-plugin，项目地址为 https://github.com/takari/takari-maven-plugin。要使用到这个功能，需要进入项目的根目录并运行以下命令：

```
mvn -N io.takari:maven:wrapper
```

还可以指定 Maven 版本号：

```
mvn -N io.takari:maven:wrapper -Dmaven=3.5.2
```

选项-N 表示－non-recursive，因此包装程序将仅应用于当前目录的主项目，而不应用于任何子模块。

之后可以使用以下命令进行程序的构建：

```
mvnw clean install
```

在这之后，Maven 将会下载到目录"$USER_HOME/.m2 /wrapper/dists"下。

> **注　意**
>
> 使用官方源进行下载将会花费很长的时间。在./mvn/wrapper/maven- wrapper.properties 文件中将配置项 distributionUrl 设置为阿里云的分发源即可解决，例如：
>
> ```
> distributionUrl=https://maven.aliyun.com/repository/central/org/apache/maven/apache-maven/3.6.3/apache-maven-3.6.3-bin.zip
> ```

1.5 构建第一个 Spring Boot 项目

本节将介绍如何从零开始构建一个 Spring Boot 项目。主要需求是使用 MVC 模式配合 Mustache 模板（一种模板引擎）实现一个简单的博客程序，为这个程序编写若干个测试用例。最后还需要发布博客相关的数个 API，让该程序除了是一个"单体应用"之外，还可以成为一个前后端分离架构中的"Web 后端程序"。

1.5.1 经典"Hello World"

按照 1.3.4 小节中介绍的 Spring Boot CLI 初始化一个 Spring Boot 项目。

【示例 1-1】

我们最先能预想到的依赖有 web、mustache、jpa、devtools，根目录暂且设定为 com.example.blog，将项目目录设置为 myblog。输入以下指令：

```
spring init -dweb,mustache,jpa,h2,devtools --package-name=com.example.blog myblog
```

创建好项目之后，编写第一个 Controller 类。还记得"最佳实践"的工程结构吗？参照那个结构创建目录 com.example.myblog.controller。在该目录下创建"Html-Controller"类。

```java
package com.example.myblog.controller;

import org.springframework.stereotype.Controller;
import org.springframework.ui.Model;
import org.springframework.web.bind.annotation.GetMapping;

@Controller
public class HtmlController {

    @GetMapping("/")
    public String blog(Model model) {
        model.addAttribute("title", "Hello World");
        return "blog";
    }
}
```

在创建好 HtmlController 之后，还需要创建一个对应的模板文件，以展示 Controller 返回的内容。在"resource/templates"目录下创建模板文件"blog.mustache"以及一些子模板。当然，这个名字可以是任何"*.mustache"，只要 controller 中的字符串返回值与文件名保持一致即可。

在路径 src/main/resources/templates/下创建 blog.mustache：

```
{{> header}}

<h1>{{title}}</h1>
```

```
{{> footer}}
```

{{> header}}为子模板，这些类似的子模板在运行时才会呈现。

在路径 src/main/resources/templates/下创建 header.mustache：

```
<html>
<head>
  <title>{{title}}</title>
</head>
<body>
```

resources/templates/footer.mustache：

```
</body>
</html>
```

在以上文件准备好了之后，通过 mvn spring-boot:run 或者其他方式来运行这个程序。在浏览器中输入：http://localhost:8080/，将会看到一个大大的"Hello World"。

1.5.2　使用 JUnit 5 测试

在完成了一个基础的功能之后，为了保证功能的正确性，需要针对该功能编写若干测试用例。自动化的测试在整个开发流程中是不可或缺的。尽可能多地编写测试用例，不仅可以帮助开发人员减少 Bug，还能促进开发人员对需求的理解。这里需要使用一款经典的测试框架 JUnit 5 来帮助实现对程序的自动化测试。

【示例 1-2】

在路径 src/test/java/com/example/myblog 下创建 IntegrationTests.java：

```
@SpringBootTest(classes = {BlogApplication.class}, webEnvironment =
SpringBootTest.WebEnvironment.RANDOM_PORT)
class IntegrationTests{

    //Spring 提供的在测试环境下访问 Rest 服务的客户端
    @Autowired
    TestRestTemplate restTemplate;

    @Test
    void assertBlogPageTitle_Content_And_StatusCode() {
        //访问路径"/"，以 String 类型来解析响应的主体 entity.body
        ResponseEntity<String> entity = restTemplate.getForEntity("/",
String.class);
        //判断响应的状态码为 HttpStatus.OK，即 200
        assertThat(entity.getStatusCode()).isEqualTo(HttpStatus.OK);
        //判断 entity.body 包含"<h1>Hello World</h1>"
        assertThat(entity.getBody()).contains("<h1>Hello World</h1>");
    }
}
```

这是一个借助注解"@SpringBootTest"实现的集成测试的测试用例。方法的命名结合了驼峰命名和蛇形命名，在方法名过长的情况下有助于提高可读性。在使用 JUnit 5 编写测试用例时，通常会把编写的测试用例分为单元测试和集成测试。这两者的区别在于测试的粒度不同。单元测试的粒度通常测试一个独立的模块甚至单独的一个类。测试过程中，该模块不与依赖项进行任何交互，以确认该模块内部在做正确的事。集成测试的粒度则覆盖多个模块，关注多个模块在协同工作的情况下是否正常。

目前需要测试的内容是控制器、模板文件在 Spring Boot 的协同工作结果，这是一个集成测试，所以需要用到 Spring Boot 提供的"@SpringBootTest"注解在测试环境创建应用程序上下文。

有时候需要在给定类之前或者之后执行一个方法，此时需要依赖到注解"@BeforeAll"和"@AfterAll"。不过，如果把这两个注解在 JUnit 5 中用于修饰常规方法，那么在使用之前需要编写一个配置文件。因为在默认情况下，JUnit 5 要求注解"@BeforeAll"和"@AfterAll"修饰的对象为静态方法。

在路径 src/test/resources 下创建配置文件 junit-platform.properties：

```
junit.jupiter.testinstance.lifecycle.default = per_class
```

配置完成之后，就可以再更新上文的 IntegrationTests.java 了：

```java
@SpringBootTest(classes = {BlogApplication.class}, webEnvironment = SpringBootTest.WebEnvironment.RANDOM_PORT)
class IntegrationTests{

    //Spring 提供的在测试环境下访问 Rest 服务的客户端
    @Autowired
    TestRestTemplate restTemplate;

    @BeforeAll
    void setup() {
        System.out.println(">> Setup");
    }

    @Test
    void assertBlogPageTitle_Content_And_StatusCode() {
        //访问路径"/"，以 String 类型来解析响应的主体 entity.body
        ResponseEntity<String> entity = restTemplate.getForEntity("/", String.class);
        //判断响应的状态码为 HttpStatus.OK，即 200
        assertThat(entity.getStatusCode()).isEqualTo(HttpStatus.OK);
        //判断 entity.body 包含"<h1>Hello World</h1>"
        assertThat(entity.getBody()).contains("<h1>Hello World</h1>");
    }

    @Test
    void assertArticlePageTitle_Content_And_StatusCode() {
        System.out.println(">> TODO");
    }

    @AfterAll
    void teardown() {
        System.out.println(">> Tear down");
    }
}
```

1.5.3 创建工具类 CommonUtil

在构建博客应用的过程中，需要用到一些工具方法。目前需要的工具方法有：格式化时间和将标题转换为 slug 格式（slug 格式是一种方便构建 URI 命名方式的格式）。例如有一篇文章标题为 "This is a title"，那么人们期望的 URI 格式可能为：

```
www.example.com/article/This is a title
```

但事实上这并不是一个有效的 URI。为了成为有效的 URI，空格会被转义为%20：

```
www.example.com/article/This%20is%20a%20title
```

不得不说，转义后的标题可读性变差了许多。为了增强 URI 的可读性，这里的空格可以被替换为 "-" 字符：

```
www.example.com/article/This-is-a-title
```

这种格式就是 slug 格式，在定义 API 的时候推荐使用该格式来命名超过一个单词的路径名。

【示例 1-3】

在路径 src/main/java/com/example/myblog/util 之下创建 CommonUtil.java：

```java
public class CommonUtil {

    private static final DateTimeFormatter englishDateFormatter;
    private static final Map<Long, String> daysLookup;

    public static String format(LocalDateTime localDateTime) {
        return localDateTime.format(englishDateFormatter);
    }

    //将 title 转换为 slug 格式："this is a title" -> "this-is-a-title"
    public static String toSlug(String title) {
        return String.join("-", title.toLowerCase()
                .replace("\n", " ")
                .replace("[^a-z\\d\\s]", " ")
                .split(" "))
                .replace("-+", "-");
    }

    static {
        daysLookup = buildDaysLookup();
        englishDateFormatter = new DateTimeFormatterBuilder()
                .appendPattern("yyyy-MM-dd")
                .appendLiteral(" ")
                .appendText(ChronoField.DAY_OF_MONTH, daysLookup)
                .appendLiteral(" ")
                .appendPattern("yyyy")
                .toFormatter(Locale.ENGLISH);
    }

    //创建 appendText 需要用到的参数 daysLookup，用于提供 appendText 时的映射关系
```

```
    private static Map<Long, String> buildDaysLookup() {
        Map<Long, String> ret = new HashMap<>();
        for (int i = 1; i <= 31; i++) {
            ret.put((long) i, getDayOfMonthSuffix(i));
        }
        return ret;
    }

    //根据天数的个位获得对应的后缀
    private static String getDayOfMonthSuffix(int n) {
        switch (n % 10) {
            case 1:
                return "${n}st";
            case 2:
                return "${n}nd";
            case 3:
                return "${n}rd";
            default:
                return "${n}th";
        }
    }
}
```

1.5.4　使用 JPA 进行数据持久化

这里计划使用 JPA（Java Persistence API）配合 H2 数据库进行数据的持久化，将文章以及作者信息写入到数据库中，并提供从数据库中读取对应数据的方法。当然，使用另外的数据库作为数据源也是可以的。

【示例 1-4】

首先需要创建文章以及作者的实体类。在路径 src/main/java/com/example/myblog/entity 下创建 Article.java：

```
@Entity
public class Article {

    @Id
    @GeneratedValue
    //主键
    private Long id;
    //标题
    private String title;
    //摘要
    private String headline;
    //内容
    private String content;
    @ManyToOne
    //作者
    private User author;
    //slug 格式的标题
    private String slug;
    //创建时间
```

```java
    private LocalDateTime addedAt;

    public Article() {
        addedAt = LocalDateTime.now();
    }

    public Long getId() {
        return id;
    }

    //chain 风格的 setter
    public Article setId(Long id) {
        this.id = id;
        return this;
    }
    //以下为除 id 字段各个字段的 getter 与 setter,在此省略……
}
```

在路径 src/main/java/com/example/myblog/entity 下创建 User.java:

```java
@Entity
public class User {
    @Id
    @GeneratedValue
    //主键
    private Long id;
    //登录名
    private String login;
    //名字
    private String firstName;
    //姓氏
    private String lastName;
    //描述
    private String description;
    //省略了 getter 与 setter 方法
}
```

以上两个类中包含 JPA 中提供的几个注解:

- @Entity:表明所修饰的类是一个实体类,如 Article.java。其默认的数据库表名为"article",也可以通过注解的 name 属性修改表名,如@Entity(name="article_alias")。
- @Id:指定字段为主键。
- @GeneratedValue:与@Id 配合使用,指定字段的生成策略为通用的生成策略。
- @ManyToOne:声明该字段与对应的对象存在一对多的关系。

创建完实体类之后,需要根据需求创建对应的 repository 类,以支持对数据库的增删改查(CURD)。在路径 src/main/java/com/example/myblog/ repository 下创建 ArticleRepository.java:

```java
public interface ArticleRepository extends CrudRepository<Article, Long> {
    //通过 slug 找到对应的文章
    Article findBySlug(String slug);
    //查询所有的文章并通过添加时间以及描述进行排序
    Iterable<Article> findAllByOrderByAddedAtDesc();
}
```

在路径 src/main/java/com/example/myblog/repository 下创建 UserRepository.java：

```java
public interface UserRepository extends CrudRepository<User, Long> {
    //通过登录名找到对应的用户
    User findByLogin(String login);
}
```

在创建 Repository 类的时候，仅需要创建一个继承 CrudRepository<T, ID> 的 interface 接口即可。泛型标记符 T 为实体类的类型，ID 为实体主键的类型。创建好对应的 interface 接口之后，剩余的工作将交给 JPA 自动完成。或许这样会带来疑惑，对数据库增删改查就不用编写 SQL 语句吗？在这里，JPA 遵循"约定优先于配置"原则，只要根据 Spring 以及 JPQL 的约定来对方法进行命名，对应 SQL 语句的生成就都交给框架来实现了。具体的 JPA 使用技巧在本书第 4 章会展开介绍。

实现了数据持久化之后，免不了对功能做一番测试。在路径 src/test/java/com/exa-mple/myblog 下创建 RepositoriesTests.java：

```java
@DataJpaTest
public class RepositoriesTests {

    @Autowired
    TestEntityManager entityManager;

    @Autowired
    UserRepository userRepository;

    @Autowired
    ArticleRepository articleRepository;

    @Test
    void whenFindByIdOrNull_thenReturnArticle() {
        //创建一个用户对象
        User leili = new User().setLogin("leili").setFirstName("Lei").setLastName("Li");
        //将用户对象转变为托管状态
        entityManager.persist(leili);
        //创建一个文章对象
        Article article = new Article().setTitle("Spring Framework 5.0 goes GA").setHeadline("Dear Spring community ...")
                .setContent("Lorem ipsum").setAuthor(leili);
        //将文章对象转变为托管状态
        entityManager.persist(article);
        //将托管状态的对象写入到数据库
        entityManager.flush();
        //根据 Id 查询文章否则返回 null
        Article found = articleRepository.findById(article.getId()).orElse(null);
        //断言保存前的对象与返回值相等
        assertThat(article).isEqualTo(found);
    }

    @Test
    void whenFindByLogin_thenReturnUser() {
        //创建一个用户对象
        User leili = new
```

```
User().setLogin("leili").setFirstName("Lei").setLastName("Li");
        //将对象直接保存至数据库
        entityManager.persistAndFlush(leili);
        //根据登录名查询对象
        User found = userRepository.findByLogin(leili.getLogin());
        //断言找到的对象与保存前的对象相等
        assertThat(found).isEqualTo(leili);
    }
}
```

在这段测试代码中，可以看到新的注解"@DataJpaTest"。该注解是 Spring Boot 为 JPA 组件测试而提供的支持。使用该注解之后，将会把全局的自动配置禁用，转而开启仅与 JPA 测试有关的配置。并且带有@DataJpaTest 注解的测试都是事务性的，这意味着每个测试结束之后都会进行回滚，避免数据库中数据被测试所影响。

1.5.5 修改控制器以及对应模板文件

对数据库的增删改查已经通过 JPA 实现了，接下来修改之前实现的控制器以及模板文件。

更新 blog.mustache：

```
{{> header}}

<h1>{{title}}</h1>

<div class="articles">

    {{#articles}}
        <section>
            <header class="article-header">
                <h2 class="article-title"><a
href="/article/{{slug}}">{{title}}</a></h2>
                <div class="article-meta">By
<strong>{{author.firstName}}</strong>, on <str
    ong>{{addedAt}}</strong></div>
            </header>
            <div class="article-description">
                {{headline}}
            </div>
        </section>
    {{/articles}}
</div>

{{> footer}}
```

在路径 src/main/resources/templates/下创建 article.mustache：

```
{{> header}}

<section class="article">
    <header class="article-header">
        <h1 class="article-title">{{article.title}}</h1>
        <p class="article-meta">By
```

```
<strong>{{article.author.firstName}}</strong>, on <strong>{{art
   icle.addedAt}}</strong></p>
    </header>

    <div class="article-description">
        {{article.headline}}

        {{article.content}}
    </div>
</section>

{{> footer}}
```

更新完模板文件之后，要着手更新对应的控制器了。在此之前需要创建一个 RenderedArticle.java。因为在很多情况下，页面展示的内容并不完全是数据库中存储的内容。这时需要借助业务对象模型（也叫领域模型）来帮助实体进行转换。Rendere-dArticle.java 便是领域模型中的 VO（view object）表现层对象。

在路径 src/main/java/com/example/myblog/domain 下创建 RenderedArticle.java：

```java
public class RenderedArticle {
    //slug 格式的标题
    private String slug;
    //标题
    private String title;
    //摘要
    private String headline;
    //内容
    private String content;
    //作者
    private User author;
    //创建时间
    private String addedAt;
    //省略了 getter 与 setter
}
```

修改 HtmlController.java：

```java
@Controller
public class HtmlController {

    private final ArticleRepository repository;

    public HtmlController(ArticleRepository repository) {
        this.repository = repository;
    }

    @GetMapping("/")
    public String blog(Model model) {
        //返回 title 字符串
        model.addAttribute("title", "Blog");
        //返回 articles 列表
        //这里用到了 StreamSupport.stream() 方法，将 repository.
findAllByOrderByAddedAtDesc() 的结果
        //转换为 stream，依次调用 render 方法进行对象的类型转换
```

```
            model.addAttribute("articles",
StreamSupport.stream(repository.findAllByOrderByAddedAtDesc().spliterator(),
true)
                    .map(this::render)
                    .collect(Collectors.toList()));
            return "blog";
        }

        //根据 slug 查询文章
        @GetMapping("/article/{slug}")
        public String article(@PathVariable String slug, Model model) {
            Article article = repository.findBySlug(slug);
            if (article == null) {
                throw new ResponseStatusException(HttpStatus.NOT_FOUND, "This article does not exist");
            }
            RenderedArticle renderedArticle = render(article);
            //返回 title 对象
            model.addAttribute("title", renderedArticle.getTitle());
            //返回 article 对象
            model.addAttribute("article", renderedArticle);
            return "article";
        }

        //对象转换
        private RenderedArticle render(Article article) {
            return new RenderedArticle()
                    .setTitle(article.getTitle())
                    .setHeadline(article.getHeadline())
                    .setSlug(article.getSlug())
                    .setContent(article.getContent())
                    .setAuthor(article.getAuthor())
                    .setAddedAt(CommonUtil.format(article.getAddedAt()));
        }
    }
```

还缺少数据的初始化，可以通过 ApplicationRunner 的方式进行配置。在程序启动前，调用数据保存的方法，实现数据的初始化。

在路径 src/main/java/com/example/myblog/config 下创建 BlogConfiguration.java：

```
    @Configuration
    public class BlogConfiguration implements ApplicationRunner{

        private final UserRepository userRepository;
        private final ArticleRepository articleRepository;

        public BlogConfiguration(UserRepository userRepository, ArticleRepository articleRepository) {
            this.userRepository = userRepository;
            this.articleRepository = articleRepository;
        }

        @Override
        public void run(ApplicationArguments args) throws Exception {
            //依次保存一条作者信息和两条文章信息
```

```java
        User meimeihan= userRepository.save(new User()
                .setLogin("meimeihan ")
                .setFirstName("meimei")
                .setLastName("han"));
        articleRepository.save(new Article()
                .setTitle("the title1")
                .setHeadline("headline1")
                .setContent("content1")
                .setAuthor(meimeihan));
        articleRepository.save(new Article()
                .setTitle("the title2")
                .setHeadline("headline2")
                .setContent("content2")
                .setAuthor(meimeihan));
    }
}
```

同样,再编写一条测试用例来测试一下。更新 IntegrationTest.java:

```java
@SpringBootTest(classes = {BlogApplication.class}, webEnvironment =
SpringBootTest.Web
    Environment.RANDOM_PORT)
    class IntegrationTest {

    @Autowired
    TestRestTemplate restTemplate;


    @BeforeAll
    void setup() {
        System.out.println(">> Setup");
    }

    @Test
    void assertBlogPageTitle_Content_And_StatusCode() {
        System.out.println(">> Assert blog page title, content and status code");
        ResponseEntity<String> entity = restTemplate.getForEntity("/", String.class);
        assertThat(entity.getStatusCode()).isEqualTo(HttpStatus.OK);
        assertThat(entity.getBody()).contains("<h1>Blog</h1>", "title1");
    }

    @Test
    void assertArticlePageTitle_Content_And_StatusCode() {
        System.out.println(">> Assert article page title, content and status code");
        String title = "title1";
        ResponseEntity<String> entity =
restTemplate.getForEntity(String.format("/article/%s",
CommonUtil.toSlug(title)), String.class);
        assertThat(entity.getStatusCode()).isEqualTo(HttpStatus.OK);
        assertThat(entity.getBody()).contains(title, "headline1", "content1");
    }
```

```
    @AfterAll
    void teardown() {
        System.out.println(">> Tear down");
    }
}
```

启动应用,跳转到地址 http://localhost:8080/。看到可单击链接的文章列表,在单击后能查看文章了吗?如果成功了的话,就说明以上用例已经精准无误地完成了。

1.5.6 发布 HTTP 接口

现在大多 Web App 都有跨平台访问的需求。为了满足这一需求,前后端分离的架构渐渐成为主流。在前后端分离的架构中,大多选用 RESTful 风格的 Web API。这里将借助注解"@RestController"来实现一个 Web API 服务,用以查询 article 以及 user 的信息。

【示例 1-5】

在路径 src/main/java/com/example/myblog/controller 下创建 ArticleController.java:

```
@RestController
@RequestMapping("/api/article")
public class ArticleController {

    private final ArticleRepository articleRepository;

    public ArticleController(ArticleRepository articleRepository) {
        this.articleRepository = articleRepository;
    }

    @GetMapping("/")
    public Iterable<Article> findAll() {
        return articleRepository.findAllByOrderByAddedAtDesc();
    }

    @GetMapping("/{slug}")
    public Article findOne(@PathVariable String slug) throws
HttpStatusCodeException {
        Article result = articleRepository.findBySlug(slug);
        //当查询结果为 null,返回 404
        if (result == null) {
            throw new ResponseStatusException(HttpStatus.NOT_FOUND, "This
article dose not exit");
        }
        return result;
    }
}
```

在路径 src/main/java/com/example/myblog/controller 下创建 UserController.java:

```
@RestController
@RequestMapping("/api/user")
```

```java
public class UserController {

    private final UserRepository userRepository;

    public UserController(UserRepository userRepository) {
        this.userRepository = userRepository;
    }

    @GetMapping("/")
    public Iterable<User> findAll() {
        return userRepository.findAll();
    }

    @GetMapping("/{login}")
    public User findOne(@PathVariable String login) {
        User result = userRepository.findByLogin(login);
        //当查询结果为 null, 返回 404
        if (result == null) {
            throw new ResponseStatusException(HttpStatus.NOT_FOUND, "This user does not exit");
        }
        return result;
    }
}
```

实现了 Rest 服务对应的控制器之后，需要编写与之对应的测试用例。对控制器层的测试有些类似于对 JPA 的测试。下文将通过 Mockito（mocking 框架）对 DAO 层进行模拟，以消除 DAO 层的返回结果对测试结果的影响，最终达到仅校验 controller 层的目的。

在路径 src/test/java/com/example/myblog 下创建 HttpControllersTests.java：

```java
@WebMvcTest
public class HttpControllersTests {

    @Autowired
    private MockMvc mockMvc;

    @MockBean
    private UserRepository userRepository;

    @MockBean
    private ArticleRepository articleRepository;

    @Test
    void listArticles() throws Exception {
        User leili = new User().setLogin("leili").setFirstName("lei").setLastName("li");
        Article article1 = new Article().setTitle("Spring Framework 5.0 goes GA")
            .setHeadline("headline1").setContent("content1")
                .setAuthor(leili);
        Article article2 = new Article().setTitle("Spring Framework 4.3 goes GA")
            .setHeadline("headline2").setContent("content2")
                .setAuthor(leili);
```

```
            //模拟 articleRepository.findAllByOrderByAddedAtDesc()方法。当调用该方法
时，返回 article1 与 article2 组成的列表
when(articleRepository.findAllByOrderByAddedAtDesc()).thenReturn(Arrays.asList
(article1, article2));
            //调用接口"/api/article/"
            mockMvc.perform(MockMvcRequestBuilders.get("/api/article/")
    .accept(MediaType.APPLICATION_JSON))
                //期望调用结果的 statuscode 为 2xx
                .andExpect(status().is2xxSuccessful())
                //期望内容类型为 APPLICATION_ JSON
                .andExpect(content().contentType(MediaType.APPLICATION_JSON))
                .andExpect(jsonPath("$.[0].author.login").value(leili.getLogi
n()))
                .andExpect(jsonPath("$.[0].slug").value(article1.getSlug()))
                .andExpect(jsonPath("$.[1].author.login").value(leili.getLogi
n()))
                .andExpect(jsonPath("$.[1].slug").value(article2.getSlug()));
    }
}
```

1.6 Spring Boot 自动配置与外部配置

在之前的章节反复提及 Spring Boot 的"自动配置"，那么自动配置大概在做什么？如果程序有更多的自定义配置的需求，在 Spring Boot 开发的过程中又需要做什么？本节会对这些内容做一个基本介绍。

1.6.1 自动配置

有读者可能会问：自动配置，自动在哪？答案并不复杂。Spring Boot 的自动配置会在没有自定义配置的情况下，尝试根据现有的 jar 包依赖对程序进行配置。以 1.5 节中实现的程序为例，在该程序的实现过程中，这里并没有直接地编写配置文件就使用上了 H2 数据库。原因就在于 Pom.xml 中声明了 H2 依赖项，并且没有其他数据源被人为添加到配置文件当中。Spring Boot 在这种场景下，默认将 H2 配置为该项目的数据库。

在不做任何配置的情况下启动项目。输入以下命令以使用 Spring-Boot Maven 插件运行 Spring Boot 项目：

```
mvn spring-boot:run
```

程序成功启动之后，终端程序上将打印如下日志信息：

```
Using dialect: org.hibernate.dialect.H2Dialect
```

看到了这行日志，说明程序已自动将 H2 数据库作为默认数据源配置在项目中了。

1.6.2 外部配置

自动配置能简化开发流程，但在真实的开发情景下还是需要开发人员手动进行配置才能满足多种多样的需求。Spring Boot 提供了多种途径的外部配置来协助开发人员，以便于在不同环境能使用相同的代码。使用 properties 文件、yaml 文件、环境变量和命令行参数，结合@Value 与@ConfigurationProperties 等注解，将配置的值注入到程序当中。

为了将外部配置和自动配置配合使用，Spring Boot 为这些配置的加载指定了一个顺序：

（1）Devtools 全局配置文件，路径：~/.spring-boot-devtools.properties（当 Spring Boot Devtools 启用）。

（2）测试用例中的"@TestPropertySource"注解。

（3）测试用例中的"properties"属性。在@SpringBootTest 和测试注解上可用，用于测试应用程序的特定部分。

（4）命令行参数。

（5）环境变量"SPRING_APPLICATION_JSON"中的字段。

（6）"ServletConfig"初始化参数。

（7）"ServletContext"初始化参数。

（8）"java:comp/env"中的 JNDI 参数。

（9）Java 系统属性(System.getProperties())。

（10）系统环境变量。

（11）配置文件"RandomValuePropertySource"，该配置文件都是随机的（random.*）。

（12）打包文件外包含于特定 profile 的配置文件（application-{profile}.propertie-s/yaml）。

（13）打包文件内包含于特定 profile 的配置文件（application-{profile}.propertie-s/yaml）。

（14）打包文件外的配置文件（application.properties/yaml）。

（15）打包文件内的配置文件（application.properties/yaml）。

（16）配置在@Configuration 配置文件中的"@PropertySource"注解。

（17）默认配置。通过 SpringApplication.setDefaultProperties()配置的属性。

1.6.3 命令行配置

在上文的介绍中可以了解到，命令行配置的优先级相对来说还是比较高的。通过"--"符号声明配置项，比如--server.port=9000 可以配置服务的端口。在 IDEA 的配置项中，打开启动项配置页"Run/Debug Configuration"，在 Program arguments 栏将配置内容输入即可在启动时附上配置项，如图 1.14 所示。

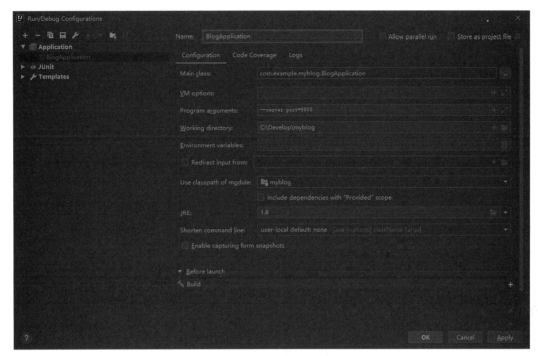

图 1.14　启动配置页

1.6.4　application.yaml/properties 配置文件

application.yaml/properties 配置文件是相对来说更常用的配置方式，配合各类注解可以满足不同需求的场景。application 配置文件支持三种格式：yaml、yml 和 properties，优先级分别为 properties→yml→yaml。在配置文件内部的优先级为自下而上进行覆盖。虽然 Spring Boot 有这样的机制，但不建议创建多个不同格式的 application 配置文件，并且配置文件内的配置项务必保持唯一，否则项目的维护将是一场灾难。

推荐使用 yaml 格式的 application.yml 文件进行配置。相较 application.properties 来说，application.yml 采用的 yaml 格式层级关系更为鲜明，使用起来更为方便。

application.properties 样例：

```
server.port=8080

logging.path=/var/logs
logging.file=myapp.log
logging.config=# 此行填入配置文件路径 (default classpath:logback.xml)
logging.level.*=# 此行填入 loggers 等级, e.g.
"logging.level.org.springframework=DEBUG" (TRACE, DEBUG, INFO, WARN, ERROR, FATAL, OFF)
```

application.yml 样例：

```
server:
  port: 8080

logging:
```

```
    path: /var/logs
    file: myapp.log
    config:    # 此行填入配置文件路径 (default classpath:logback.xml)
    level.*:   # 此行填入 loggers 等级, e.g.
"logging.level.org.springframework=DEBUG" (TRACE, DEBUG, INFO, WARN, ERROR, FATAL,
OFF)
```

结合@ConfigurationProperties 注解可以将对程序内变量的赋值提取到配置文件中。下文把在 1.5 节实现的程序中做进一步的配置，将博客的主题以及 banner 提取到配置文件中。在路径 src/main/java/com/example/myblog/config 下创建 BlogProperties.java：

```
@ConfigurationProperties("blog")
public class BlogProperties {

    private String title;
    private Banner banner;
    //省略了字段的 getter 与 setter
}
```

在 BlogApplication 中启用该配置：

```
@SpringBootApplication
@EnableConfigurationProperties(BlogProperties.class)
public class BlogApplication {
    //…
}
```

在路径 src/main/resources 路径下删除初始化的 application.properties 文件，并创建 application.yml 以替代它：

```
blog:
  title: Blog
  banner:
    title: Surprise!
    content: You made it!!!
```

修改模板以及控制器，以用上这些配置属性。blog.mustache：

```
{{> header}}

<div class="articles">

    {{#banner.title}}
        <section>
            <header class="banner">
                <h2 class="banner-title">{{banner.title}}</h2>
            </header>
            <div class="banner-content">
                {{banner.content}}
            </div>
        </section>
    {{/banner.title}}

    {{#articles}}
        <section>
            <header class="article-header">
```

```
              <h2 class="article-title"><a
href="/article/{{slug}}">{{title}}</a></h2>
              <div class="article-meta">By
<strong>{{author.firstName}}</strong>, on <strong>{{addedAt}}</strong></div>
          </header>
          <div class="article-description">
              {{headline}}
          </div>
      </section>
    {{/articles}}
  </div>

  {{> footer}}
```

HtmlController.java：

```
@Controller
public class HtmlController {
    //…
    @GetMapping("/")
    public String blog(Model model) {
        model.addAttribute("title", blogProperties.getTitle());
        model.addAttribute("banner", blogProperties.getBanner());
        model.addAttribute("articles", StreamSupport.stream(repository.findAllByOrderByAddedAtDesc().spliterator(), true)
                .map(this::render)
                .collect(Collectors.toList()));
        return "blog";
    }
    //……
}
```

重启应用后，可以看到新的主页标题和内容了。

第 2 章

使用 Spring Boot 构建 Web 应用程序

在 1.5 节着手实现了第一个 Spring Boot 应用程序，大家在实现的过程中已经领略到 Spring Boot 的强大与便捷。但是，其中接触到的大部分知识因为开篇的缘故一笔带过了，本章将结合之前的示例进一步展开说明这部分内容。

本章主要涉及的知识点有：

- 数据持久化
- MVC 架构的 Web 应用
- 对多媒体数据的处理
- 拦截器与过滤器
- Spring Boot 事件
- Spring Boot 日志

2.1 实体与数据持久化

Web 应用程序终归是离不开数据的。数据在其中好比是泥土，一款 Web 应用程序构建出来的目的便是有机地组织数据，就像果树汲取了泥土的养分之后结出可口的果实。数据持久化技术自然就是绕不开的第一步。

2.1.1 数据持久化框架

数据持久化意味着应用程序可以将数据存储到非易失性的存储设施中，并能在其中执行检索操作。在早期的 Web 开发过程中，这一环节依赖 JDBC（Java DataBase Connectivity，Java 数据库基本连接）与 RDBMS（Relational Database Management System，关系型数据库管理系统）。如今有诸多数据持久化的方案可供选择，例如：Hibernate、MyBatis、JOOQ、Ebean 以及 JdbcTemplate。下面对其中 3 个相对主流的解决方案进行对比。

（1）Hibernate：一个开源的轻量级 ORM（Object Relational Mapping，对象关系映射）框架。通过将应用程序中的实体映射到关系型数据库，以实现 JPA 的数据持久性，进一步做到与数据库进行交互。这意味着开发人员无须编写 SQL 语句，仅仅通过操作实体以及调用框架所封装的方法即可实现增删改查（CURD），大大简化了应用程序的开发。

为了提高所构建应用的性能，Hibernate 还提供了诸多额外的功能，如缓存、延迟初始化等。其中缓存相当于一个介于应用与数据库之间的存储设施，作用是提升数据的获取速度。Hibernate 支持两级缓存，分别称为一级缓存与二级缓存。因为同样的功能，在不同数据库中所使用的关键字存在一定差异，Hibernate 为了支持多种数据库提供了对应各种类型的"数据库方言"。

（2）MyBatis：一个开源的轻量级持久化框架。与 ORM 框架不同，MyBatis 并不会在 Java 对象与数据库之间建立映射，而将 Java 方法映射到 SQL 语句。这意味着开发过程中所调用的 DAO 层方法，都需要由开发人员编写 SQL 语句以及对应的 Mapper。与 ORM 框架相比，MyBatis 非常强调 SQL 语句。这导致开发过程相对烦琐，却换来很大程度的自由度，为在特殊场景下编写 SQL 语句调优提供了便利。

MyBatis 虽然不是 ORM 框架，但同样提供了映射引擎以帮助 SQL 的执行结果映射至对象。该框架允许开发人员使用所有的数据库功能，例如存储过程、视图等。MyBatis 支持声明式**数据缓存**，一条语句被标记为可缓存后，其查询结果将被存储至缓存当中。在查询过程的开始，首先访问缓存以查询是否有缓存数据，未命中缓存则转而查询数据库，并将结果存储到缓存中以便下次查询。

（3）JdbcTemplate：由 Spring 提供的持久化解决方案。JdbcTemplate 并非框架，而是对原生 JDBC 的封装。在封装的过程中，JdbcTemplate 解决了数个 JDBC 开发面临的问题：

- 在执行查询之前与之后，开发人员均需要编写大量的模板代码，例如创建或关闭连接。
- 对数据库逻辑执行异常处理（try/catch）。
- 手动提交事务。
- 很难实现不同数据库之间的迁移。

2.1.2 什么是实体

实体（Entity）的概念来源于 ER 模型（Entity–Relationship），由 Peter Chen 提出用于数据库设计。在一个简单的关系型数据库当中，表的每一行对应一个实体类型的实例，表中每一个字段代表实例类型的一个属性。在关系型数据库当中，实体之间的关系是通过将一个实体的主键以外键形式存储在另一实体表中来实现的。

实体在数据持久化中的作用在于使用面向对象的形式表达数据库，操作对象即操作数据库。

2.1.3 浅谈 Spring Data JPA

2.1.1 小节介绍 Hibernate 并提到了 JPA。JPA（Java Persistence API）是由 Sun 公司推出的一套 ORM 规范。Hibernate 之于 JPA 如同 JDBC Driver 之于 JDBC。Hibernate 是 JPA 的具体实现。Spring 为了支持在使用 JPA 的场景下更方便地编码，推出了 Spring Data JPA 这一项目。Spring Data JPA 是对 JPA 的封装。

【示例 2-1】

在 1.5.4 小节我们已初步接触过 Spring Data JPA。对数据库的操作正是依赖 Spring Data JPA 的 CrudRepository。CrudRepository 提供了一组与 CRUD 有关的基础方法：

```java
public interface CrudRepository<T, ID> extends Repository<T, ID> {
    //添加一条数据
    <S extends T> S save(S entity);
    //添加多条数据
    <S extends T> Iterable<S> saveAll(Iterable<S> entities);
    //根据 id 查询一条数据
    Optional<T> findById(ID id);
    //根据 id 判断是否存在
    boolean existsById(ID id);
    //查询所有的数据
    Iterable<T> findAll();
    //根据 id 查询集合
    Iterable<T> findAllById(Iterable<ID> ids);
    //查询数据的条数
    long count();
    //根据 id 删除数据
    void deleteById(ID id);
    //根据给定的数据进行删除
    void delete(T entity);
    //根据给定的集合进行删除
    void deleteAll(Iterable<? extends T> entities);
    //删除所有数据
    void deleteAll();
}
```

使用以上提供的方法，为 1.5.4 小节中声明的 UserRepository 编写一个 DataJpaTest，以熟悉这些方法的使用：

```java
@DataJpaTest
public class ArticleRepositoryTests {

    @Autowired
    UserRepository userRepository;

    @Test
    public void saveAUser_thenFindIt() {
        //保存一条用户数据，然后进行查询
        User leili = new User().setLogin("leili").setFirstName("Lei").setLastName("Li");
```

```java
        userRepository.save(leili);
        User found = userRepository.findById(leili.getId()).orElse(null);
        assertThat(found).isNotEqualTo(null);
        assertThat(leili).isEqualTo(found);
    }

    @Test
    public void saveUserList_thenCountThem() {
        //保存一个用户数据集合,然后查询记录的数据量
        User leili = new User().setLogin("leili").setFirstName("Lei").setLastName("Li");
        User meimeihan = new User().setLogin("meimeihan").setFirstName("Meimei")
                .setLastName("Han");
        User taolin = new User().setLogin("taolin").setFirstName("Tao").setLastName("Lin");
        User jimgreen = new User().setLogin("jimgreen").setFirstName("Jim").setLastName("Green");
        List<User> toSave = Arrays.asList(leili, meimeihan, taolin, jimgreen);
        userRepository.saveAll(toSave);
        Long countThem = userRepository.count();
        assertThat(countThem).isEqualTo(4);
    }

    @Test
    public void saveAUser_thenDeleteIt() {
        //保存一条用户数据,然后删除它
        User leili = new User().setLogin("leili").setFirstName("Lei").setLastName("Li");
        userRepository.save(leili);
        userRepository.delete(leili);
        User found = userRepository.findById(leili.getId()).orElse(null);
        assertThat(found).isEqualTo(null);
    }

    @Test
    public void saveUserList_thenDeleteThem() {
        //保存一个用户数据集合,然后删除所有记录
        User leili = new User().setLogin("leili").setFirstName("Lei").setLastName("Li");
        User meimeihan = new User().setLogin("meimeihan").setFirstName("Meimei")
                .setLastName("Han");
        User taolin = new User().setLogin("taolin").setFirstName("Tao").setLastName("Lin");
        User jimgreen = new User().setLogin("jimgreen").setFirstName("Jim").setLastName("Green");
        List<User> toSave = Arrays.asList(leili, meimeihan, taolin, jimgreen);
        userRepository.saveAll(toSave);
        userRepository.deleteAll();
        Long countThem = userRepository.count();
        assertThat(countThem).isEqualTo(0);
    }
}
```

2.1.4 使用 Lombok 简化 POJO

1.5.4 小节中定义了实体类 Article 与 User。实体类的职责在"贫血模式"（领域驱动设计中的概念）下大多很简单，即存储数据。但随着字段的增长，getter 以及 setter 的声明使得类行数变得十分膨胀。本节将推荐一款插件解决这个问题——Lombok。

（1）添加 Maven 依赖。

```
<dependency>
    <groupId>org.projectlombok</groupId>
    <artifactId>lombok</artifactId>
    <version>1.18.12</version>
    <scope>provided</scope>
</dependency>
```

（2）添加 IDE 对 Lombok 的支持。在工具栏中依次选择"File"→"Settings"，打开 IDEA 的设置页面。在设置页面的左侧选择"Plugins"，打开 IDEA 的插件管理页面。在搜索栏中输入"Lombok"，即可看到所需的插件，如图 2.1 所示。

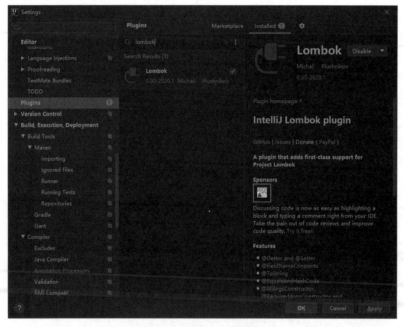

图 2.1 IDEA Lombok 插件

开启 AnnotationProcessor，使 Lombok 在编译阶段生效。操作顺序为"File"→"Settings"→"Build,Execution,Deployment"→"compiler"→"Annotation Processor"，如图 2.2 所示。

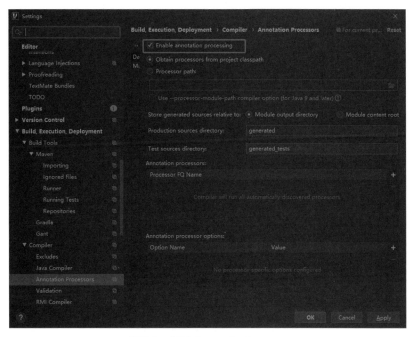

图 2.2　启用 AnnotationProcessor

（3）作用于 POJO，消除不友好的代码。使用 Lombok 注解修改 Article.java：

```java
@Accessors(chain = true)
@Setter
@Getter
@Entity
public class Article {
    @Id
    @GeneratedValue
    private Long id;
    private String title;
    private String headline;
    private String content;
    @ManyToOne
    private User author;
    private String slug;
    private LocalDateTime addedAt;

    public Article() {
        addedAt = LocalDateTime.now();
    }

    public Article setTitle(String title) {
        this.title = title;
        this.slug = CommonUtil.toSlug(title);
        return this;
    }
}
```

@Getter/@Setter 注解作用于类上用于生成所有成员变量的 getter/setter，作用于成员变量上用于生成对应的 getter/setter。@ Accessors 用于配置 getter/setter 的生成结果，当设置属性 chain 为 true

时，setter 方法将返回当前对象。

Lombok 的功能不仅限于此，更多的特性可自行参考官网。

2.2 MVC 与模板引擎

在 1.5.1 小节实现的 HelloWorld 示例中，我们已经接触到了 MVC 与模板引擎。这是一种可以迅速构建出一个 Web 应用的技术。本节将继续介绍 MVC 的相关内容，并介绍在基于 MVC 架构的开发过程中组织代码的思路。

2.2.1 MVC 架构

MVC（Model-View-Controller）是一个很经典的软件设计模式。通常用于开发 GUI（Graphical User Interface，图形用户界面）。MVC 设计模式将程序表现层逻辑分为三个模块：Model（模型）、View（视图）以及 Controller（控制器）。这种模式便于将数据的内部表达、数据的输入以及数据的输出拆分开，使代码更易组织与维护。各个模块的职责如下：

- Model：该模块在 MVC 设计模式中的职责在于维护数据的内部表达。
- View：该模块的职责在于数据的展示。对应示例中的 mustache 模板引擎。
- Controller：该模块的职责在于控制来自于 View 的输入数据，操作 Model 并将执行结果返回用于渲染 View。

三者之间的关系如图 2.3 所示。

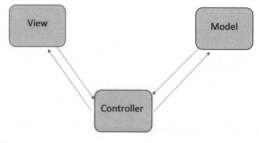

图 2.3 Model-View-Controller

使用 MVC 模式来开发软件的表现层实现了程序的高内聚与松耦合。例如 1.5.1 小节的示例中使用的 View 为 Mustache，可以在基本不修改 Model 与 Controller 的情况下，将 View 更换为 Thymeleaf 或者 Velocity。

2.2.2 Mustache 模板引擎

在之前的示例中，Mustache 已经展现出了它的威力。使用模板引擎，可以将 Web 页面的维护拆分成了动态（内容）和静态（模板）两部分。以 Mustache 的一个经典的模板举例，模板如下：

```
Hello {{name}}
You have just won {{value}} dollars!
{{#in_ca}}
Well, {{taxed_value}} dollars, after taxes.
{{/in_ca}}
```

如果传入如下内容：

```
{
  "name": "Chris",
  "value": 10000,
  "taxed_value": 10000 - (10000 * 0.4),
  "in_ca": true
}
```

那么将得到这样的结果：

```
Hello Chris
You have just won 10000 dollars!
Well, 6000.0 dollars, after taxes.
```

这样一来便可以将动态的内容插入模板，让 Web 页面根据需要生成对应的内容。

模板引擎的使用围绕标签展开，不同的标签具有不同的渲染规则。下面介绍一些常用的标签。

1. {{tag}}

该标签被称为变量（Variables），是最基本的标签类型。例如，示例"Hello {{name}}"中的 {{name}}，使用该标签之后，模板引擎将会在"内容源"中找到键为"name"对应的值——"Chris"，并将其呈现至结果当中。如果"内容源"中不包含该键值对（Key Value Pair），则不会呈现任何内容。需要注意的是，变量的内容是经过 HTML 转义的。如果需要未转义的内容，则需要三重大括号{{{data}}}。示例如下：

模板：

```
{{name}}
{{age}}
{{company}}
{{{company}}}
```

内容：

```
{
  "name": "Chris",
  "company": "<b>GitHub</b>"
}
```

结果：

```
Chris

&lt;b&gt;GitHub&lt;/b&gt;
"<b>GitHub</b>
```

2. {{#tag}}/{{/tag}}

该标签由{{#tag}}与{{/tag}}两部分组成，被称为区块（Section）。其作用在于当 tag 的内容为

false、null 或者空数组时，tag 中的内容将被隐藏。该标签功能强大，传入不同类型的内容源可以实现不同效果。

（1）当 tag 对应的值为"false"的示例如下：

模板：

```
Shown.
{{#ifShow}}
  Never shown!
{{/ ifShow }}
```

内容：

```
{
  " ifShow ": false
}
```

结果：

```
Shown.
```

（2）当 tag 对应的值为"非空数组"的示例如下：

模板：

```
{{#beatles}} {{firstName}} {{lastName}} <br/>{{/beatles}}
```

内容：

```
{
  "beatles": [{
    "firstName": "John",
    "lastName": "Lennon"
   },
   {
    "firstName": "Paul",
    "lastName": "McCartney"
   },
   {
    "firstName": "George",
    "lastName": "Harrison"
   },
   {
    "firstName": "Ringo",
    "lastName": "Starr"
   }
  ]
}
```

结果：

```
John Lennon
Paul McCartney
George Harrison
Ringo Starr
```

(3)当 tag 对应的值为"Lambda"的示例如下：

模板：

```
{{#wrapped}}
  {{name}} is awesome.
{{/wrapped}}
```

内容：

```
{
  "name": "Willy",
  "wrapped": function() {
    return function(text, render) {
      return "<b>" + render(text) + "</b>"
    }
  }
}
```

结果：

```
<b>Willy is awesome.</b>
```

(4)当 tag 对应的值为"对象"的示例如下：

模板：

```
{{#person?}}
  Hi {{name}}!
{{/person?}}
```

内容：

```
{
  "person?": { "name": "Jon" }
}
```

结果：

```
Hi Jon!
```

3. {{^tag}}/{{/tag}}

该标签被称作反区块（Inverted Section），与区块的作用恰好相反。当 tag 对应的值为 false、null 或者空数组，tag 中的内容将被展示出来。示例如下：

模板：

```
{{#repo}}
  <b>{{name}}</b>
{{/repo}}
{{^repo}}
  No repos :(
{{/repo}}
```

内容：

```
{
  "repo": []
}
```

结果：

```
No repos :(
```

4. {{> tag}}

该标签被称作部分（Partial），用于调用其他以 mustache 结尾的模板。

base.mustache：

```
<h2>Names</h2>
{{#names}}
  {{> user}}
{{/names}}
```

user.mustache：

```
<strong>{{name}}</strong>
```

结果可以当作如下单个模板：

```
<h2>Names</h2>
{{#names}}
  <strong>{{name}}</strong>
{{/names}}
```

5. {{!tag}}：

该标签为注释（Comment）。这里不再举例。

6. 设置定界符

在定界符"Delimiter"中使用等号设定自定义的定界符。示例如下：

```
{{default_tags}}
{{=<% %>=}}
<% erb_style_tags %>
<%={{ }}=%>
{{ default_tags_again }}
```

示例中分别使用了"{{=<% %>=}}"与"<%={{ }}=%>"修改了定界符，避免在模板文件使用大括号引发异常。

> **注 意**
>
> 自定义定界符不可包含空格与等号。

2.2.3 构建 MVC 架构的 Web 应用

本节将继续完善 1.5 节中所实现的应用。MVC 架构的应用往往比较注重交互，在构建应用的第一步，往往需要考虑的是页面的输入与输出。这里计划新增一个文章录入功能，需要在页面中选择作者并填写文章标题、正文等信息。

【示例 2-2】

（1）确定页面元素以及数据传输对象的结构。文章录入功能需要传输的信息有：作者名、标题、副标题、正文。根据需求设计对应的请求对象，在路径 src/main/java/ com/example/myblog/domain 下创建 SubmitArticleQuery.java：

```java
@Accessors(chain = true)
@Setter
@Getter
public class SubmitArticleQuery {
    //标题
    private String title;
    //副标题
    private String headline;
    //正文
    private String content;
    //作者名
    private String author;
}
```

（2）设计一个页面模板，满足页面输入的基本要求之外，还需要具有发起 POST 请求的功能。修改模板文件 header.mustache，引入 jquery 脚本，以便编写 ajax 请求相关的脚本：

```html
<html>
<head>
    <script src="http://code.jquery.com/jquery-1.12.4.min.js"></script>
    <title>{{title}}</title>
</head>
<body>
```

在同一目录下新增用于录入文章的模板文件 writing.mustache：

```
{{> header}}

<div class="writing">
    <b><a href="/">Home</a></b>
    <form class="to-save">
        <br> title <br>
        <input type="text" name="title">
        <br> headline <br>
        <input type="text" name="headline">
        <br> content <br>
        <textarea rows="10" cols="70" name="content"></textarea>
        <br>Author:
        <select name="author">
            {{#users}}
                <option value="{{login}}">{{login}}</option>
            {{/users}}
        </select>
        <br>
        <input type="submit">
    </form>
</div>

<script type="text/javascript">
$("form").submit(function() {
    //构造请求体
    var formObject = {};
```

```
        var formArray = $("form").serializeArray();
        $.each(formArray,
        function(i, item) {
            formObject[item.name] = item.value;
        });
        //使用 AJAX,创建 POST 请求
        $.ajax({
            type: 'POST',
            url: "/article",
            data: JSON.stringify(formObject),
            contentType: 'application/json',
            success: function(data) {
                alert(data);
            }
        });
    });
</script>

{{> footer}}
```

（3）如此一来，MVC 中的"Model"与"View"的部分已经基本实现了。开始着手创建 MVC 最后的"Controller"。该页面分别需要两个 Controller 方法来协助实现最终效果，分别用于展示页面与接收 POST 请求。更新 HtmlController.java：

```
@RequiredArgsConstructor
@Controller
public class HtmlController {

    private final ArticleRepository articleRepository;
    private final UserRepository userRepository;
    private final BlogProperties blogProperties;

    //省略若干方法

    @GetMapping("/writing")
    public String article(Model model) {
        //用于填写页面的展示
        Iterable<User> userList = userRepository.findAll();
        model.addAttribute("title", "writing");
        model.addAttribute("users", userList);
        return "writing";
    }

    @PostMapping("/article")
    @ResponseBody
    public String submitArticle(@RequestBody SubmitArticleQuery query) {
        //用于接收 POST 请求
        User author = userRepository.findByLogin(query.getAuthor());
        if (author == null) {
            //返回 400 错误码
            throw new ResponseStatusException(HttpStatus.BAD_REQUEST, "This user does not exist");
        }
        Article toSave = new Article().setAuthor(author)
                .setTitle(query.getTitle())
```

```
                .setHeadline(query.getHeadline())
                .setContent(query.getContent())
                .setSlug(CommonUtil.toSlug(query.getTitle()));
        articleRepository.save(toSave);
        return "success";
    }
}
```

submitArticle 方法仅返回类型为 String 的请求结果，而非对应名称的视图模板，因此需要在方法上方使用注解"@ResponseBody"对其进行修饰。

> **注 意**
>
> "Controller"为表现层（User Interface Layer）的重要一环，不应处理过多业务逻辑。在开发过程中若存在更多更复杂的代码逻辑，应"下沉"至业务逻辑层（Business Logic Layer）。更多内容请参考"三层架构"理论。

（4）测试环节，编写针对 submitArticle() 的测试用例。在 IntegrationTest.java 中新增一个测试方法：

```
@SpringBootTest(classes = {BlogApplication.class}, webEnvironment =
SpringBootTest.WebEnvironment.RANDOM_PORT)
class IntegrationTest {

    @Autowired
    TestRestTemplate restTemplate;
    //省略若干方法
    @Test
    void submitAnArticle() {
        System.out.println(">> Submit an article");
        SubmitArticleQuery queryFromAnonymous = new SubmitArticleQuery()
                .setAuthor("anonymous")
                .setTitle("title")
                .setHeadline("headline")
                .setContent("content");
        ResponseEntity<String> entity = restTemplate.postForEntity("/article",
queryFromAnonymous, String.class);
        //若查找不到"anonymous"这位作者，将返回错误码 400
assertThat(entity.getStatusCode()).isEqualTo(HttpStatus.BAD_REQUEST);
        SubmitArticleQuery queryFromLeili = new SubmitArticleQuery()
                .setAuthor("meimeihan")
                .setTitle("title2")
                .setHeadline("headline2")
                .setContent("content2");
        ResponseEntity<String> entity2 =
restTemplate.postForEntity("/article", queryFromLeili, String.class);
        //正常的返回结果
        assertThat(entity2.getStatusCode()).isEqualTo(HttpStatus.OK);
        assertThat(entity2.getBody()).contains("success");
    }
}
```

2.3 文件上传与下载

在 Web 应用中，对多媒体文件的操作非常常见，文件的上传与下载尤为如此。本节将通过在 Myblog 这个项目中新增图片以及附件的上传下载模块，带领读者熟悉该类型功能的开发流程。

2.3.1 文件上传

首先实现文件上传功能。当文件上传完毕之后对文件进行存储，最终返回对应的下载路径。

【示例 2-3】

（1）编写配置以指定文件的存储路径。在路径 src/main/java/ com/example/myblog/co-nfig 下创建 FileStorageProperties.java：

```
@Setter
@Getter
@ConfigurationProperties(prefix = "file")
public class FileStorageProperties {
    private String uploadDir;
}
```

（2）在 BlogApplication.java 中启用该配置：

```
@SpringBootApplication
@EnableConfigurationProperties({BlogProperties.class,
FileStorageProperties.class})
public class BlogApplication {

 public static void main(String[] args) {
  SpringApplication.run(BlogApplication.class, args);
 }
}
```

（3）在配置文件 application.yml 中添加对应配置信息，暂且将文件的存储路径设定为当前项目路径下的一个目录中：

```
file:
  upload-dir: ./assets
```

（4）编写文件存储相关业务逻辑。创建用于容纳业务逻辑代码的路径 src/main/java/com/example/myblog/service，并在此路径下创建 FileStorageService.java：

```
@Service
public class FileStorageService {

    private final Path fileStorageLocation;

    public FileStorageService(FileStorageProperties fileStorageProperties)
throws Exception {
```

```java
        this.fileStorageLocation = 
Paths.get(fileStorageProperties.getUploadDir()).toAbsolutePath().normalize();
        try {
            Files.createDirectories(this.fileStorageLocation);
        } catch (IOException e) {
            e.printStackTrace();
            throw new Exception("Could not create the directory where the uploaded files will be stored.", e);
        }
    }

    public String uploadFile(MultipartFile file) {
        //文件名
        String originalName = file.getOriginalFilename();
        //扩展名
        String extName = originalName == null || originalName.lastIndexOf(".") <= 0 ?
                null : originalName.substring(originalName.lastIndexOf("."));
        String fileName = UUID.randomUUID().toString() + extName;
        if (fileName.contains("..")) {
            throw new RuntimeException("Sorry! Filename contains invalid path sequence " + fileName);
        }
        //根据文件名获得最终的 Path 对象
        Path target = fileStorageLocation.resolve(fileName);
        try {
            //将文件流复制至目标 Path
            Files.copy(file.getInputStream(), target, StandardCopyOption.REPLACE_EXISTING);
        } catch (IOException e) {
            throw new RuntimeException("Could not store file " + fileName + ". Please try again!", e);
        }
        return fileName;
    }

    public Resource loadFile(String fileName) {
        //TODO
        return null;
    }
}
```

当前步骤提供了一个用于文件存储的方法。对存储路径执行初始化操作，当路径不存在时，则会创建该路径。之后通过将文件流复制至该路径以实现上传文件的需求。

（5）公布文件上传 API。逻辑在 Service 层实现之后，需要通过 Controller 层将对应的服务公开出来。在路径 src/main/java/com/example/myblog/controller 下创建 FileController.java：

```java
@RequiredArgsConstructor
@RestController
@RequestMapping("/file")
public class FileController {

    private final FileStorageService fileStorageService;
```

```java
    @PostMapping
    public String uploadFile(@RequestParam("file") MultipartFile file) {
        return String.format("/file/%s",
fileStorageService.uploadFile(file));
    }

    @GetMapping("/{fileName:.+}")
    public ResponseEntity<Resource> download(@PathVariable String fileName,
     HttpServletRequest request) {
        //todo
        throw new ResponseStatusException(HttpStatus.NOT_FOUND, "Under construction!");
    }
}
```

（6）至此，文件上传的功能已基本实现。接下来将要对其做一番测试，以验证代码的准确性，在路径 src/test/java/com/example/myblog 下创建 FileUploadControllerTests.java：

```java
@RunWith(SpringRunner.class)
@SpringBootTest
public class FileUploadControllerTests {

    @Autowired
    private WebApplicationContext wac;

    private MockMvc mockMvc;

    @Before
    public void setup() {
        mockMvc = MockMvcBuilders.webAppContextSetup(wac).build();
    }

    @Test
    public void whenUploadFile_thenReturnAnUrl() throws Exception {
        String result = mockMvc.perform(
                MockMvcRequestBuilders
                        .multipart("/file")
                        .file(
                                new MockMultipartFile("file",
                                        "test.txt",
                                        ",multipart/form-data",
                                        "hello upload".getBytes(StandardCharsets.UTF_8))
                        )
        ).andExpect(MockMvcResultMatchers.status().isOk())
                .andReturn().getResponse().getContentAsString();
        //断言返回结果中包含路径名"file"
        assertThat(result).contains("file");
    }

}
```

2.3.2 文件下载

在实现了文件上传功能之后，本小节将展示文件下载功能的实现过程。操作步骤如下：

【示例 2-4】

（1）在 FileStorageService.java 新增下载相关方法 loadFile，通过路径名获得文件的 Resource 对象：

```java
public Resource loadFile(String fileName) throws FileNotFoundException {
    Path filePath = fileStorageLocation.resolve(fileName).normalize();
    try {
        Resource resource = new UrlResource(filePath.toUri());
        if (resource.exists()) {
            return resource;
        } else {
            throw new FileNotFoundException("file not found" + fileName);
        }
    } catch (MalformedURLException e) {
        throw new FileNotFoundException("file not found" + fileName);
    }
}
```

（2）在 FileController.java 补全 download 方法，获取对应文件的 Resource 对象之后，将其写入响应体中：

```java
@Slf4j
@RequiredArgsConstructor
@RestController
@RequestMapping("/file")
public class FileController {

    private final FileStorageService fileStorageService;

    @PostMapping
    public String uploadFile(@RequestParam("file") MultipartFile file) {
        return String.format("/file/%s", fileStorageService.uploadFile(file));
    }

    @GetMapping("/{fileName:.+}")
    public ResponseEntity<Resource> download(@PathVariable String fileName,
     HttpServletRequest request) {
        Resource ret;
        try {
            //获取文件的 Resource 对象
            ret = fileStorageService.loadFile(fileName);
        } catch (FileNotFoundException e) {
            throw new ResponseStatusException(HttpStatus.NOT_FOUND, "File not found.");
        }
        //将结果写入响应中
        return downloadFile(ret, request);
```

```java
    }

    private ResponseEntity<Resource> downloadFile(Resource resource,
HttpServletRequest request) {
        String contentType = null;
        try {
            contentType =
request.getServletContext().getMimeType(resource.getFile().getAbsolutePath());
        } catch (IOException e) {
            log.error("Could not determine file type.");
        }
        if (contentType == null) {
            contentType = "application/octet-stream";
        }
        return
ResponseEntity.ok().contentType(MediaType.parseMediaType(contentType))
                .header(HttpHeaders.CONTENT_DISPOSITION,
"attachment;filename=\"" +
    resource.getFilename())
                .body(resource);
    }

}
```

（3）编写测试用例。先上传一个测试文件，然后通过 API 下载该文件并比对文件内容：

```java
@RunWith(SpringRunner.class)
@SpringBootTest
public class FileUploadControllerTests {

    @Autowired
    private WebApplicationContext wac;

    private MockMvc mockMvc;

    @Before
    public void setup() {
        mockMvc = MockMvcBuilders.webAppContextSetup(wac).build();
    }
    //省略若干方法
    @Test
    public void uploadFile_AndDownloadIt() throws Exception {
        String content = "hello upload";
        String downloadUrl = mockMvc.perform(
            MockMvcRequestBuilders
                    .multipart("/file")
                    .file(
                            new MockMultipartFile("file",
                                    "test.txt",
                                    ",multipart/form-data",
                                    content.getBytes(StandardCharsets.UTF_8))
                    )
        ).andExpect(MockMvcResultMatchers.status().isOk())
                .andReturn().getResponse().getContentAsString();
        MvcResult downloadResult = mockMvc.perform(
```

```
                MockMvcRequestBuilders
                        .get(downloadUrl)
                        .contentType(MediaType.APPLICATION_OCTET_STREAM)
        ).andExpect(MockMvcResultMatchers.status().isOk()).andReturn();
assertThat(downloadResult.getResponse().getContentAsString()).isEqualTo(content);
    }
}
```

2.4　Spring Boot 日志

在软件开发过程中，偶尔发生程序没有按照预期方向执行的情况，是在所难免的。为了避免程序出现 Bug 但又不能 Debug 去定位的情况，提供丰富的日志功能就是最"通用"的选择了。好消息是 Spring Boot 集成的日志功能非常强大且容易使用，本节将讲解 Spring Boot 日志相关的内容。

2.4.1　使用预设配置

自动配置在没有特殊的需求时，日志功能是开箱即用的状态。在路径 src/main/java/com/example/myblog/controller 下创建 LogController.java：

【示例 2-5】

```
import org.slf4j.Logger;
import org.slf4j.LoggerFactory;
import org.springframework.web.bind.annotation.GetMapping;
import org.springframework.web.bind.annotation.RequestMapping;
import org.springframework.web.bind.annotation.RestController;

@RequestMapping("/log")
@RestController
public class LogController {

    Logger logger = LoggerFactory.getLogger(LogController.class);

    @GetMapping
    public String justShowSomeLog() {
        logger.trace("TRACE Message.");
        logger.debug("DEBUG Message.");
        logger.info("INFO Message.");
        logger.warn("WARN Message.");
        logger.error("ERROR Message.");
        return "Bro,go check your console.";
    }
}
```

访问 http://localhost:8080/log 并查看控制台，将会看到有日志信息在不断输出。

2.4.2 基础配置

以上示例实现了之后，细心的读者可能会疑惑：为什么代码中编写了五条日志，但控制台中的记录只输出了三条，如图 2.4 所示。

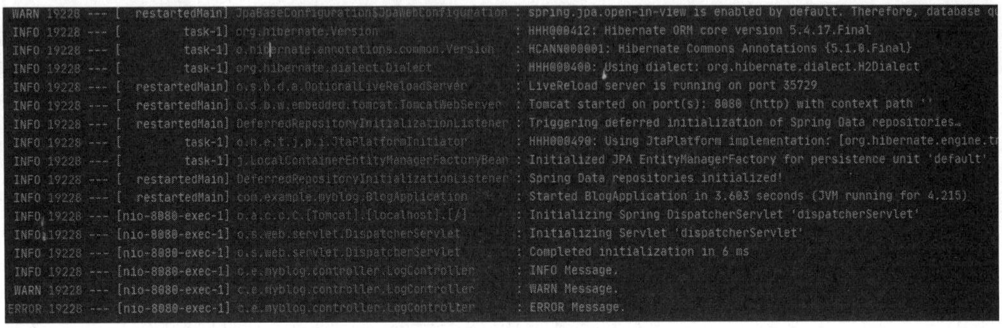

图 2.4 默认日志等级

原因在于这五条日志从上至下分别对应五种日志等级：跟踪（Trace）、调试（Debug）、信息（Info）、警告（Warn）和错误（Error）。其中 Info 为默认的日志等级，代表着仅显示 Info、Warn、Error 这三个等级的日志。如果有改变日志等级的需求，需要做一些基本的配置工作。

以将默认日志等级调整至 Debug 为例，在 application.yml 中新增一条配置：

```
logging:
  level:
    root: debug
```

这条配置的含义为：设置 root 级别，即所有包内的日志等级为 debug。重新启动项目并访问 http://localhost:8080/log，可以观察到增加了许多不同的日志信息，如图 2.5 所示。

图 2.5 root 日志等级设置为 Debug

该配置项支持以包为单位的日志等级控制，这意味着可以将 Spring Boot 以及其他各种依赖的日志等级提升至 Warn，业务代码的日志等级下调至 Debug：

```
logging:
  level:
    root: warn
com.example.myblog: debug
```

结果如图 2.6 所示。

图 2.6 自定义日志等级

除了以上所介绍的 logging.level 配置之外，Spring Boot 还提供了另外一些配置项以改变日志的行为：

- logging.file：指定日志输出的目标文件。为避免用户使用中产生不必要的困惑，该属性在 2.2.x 及以上版本被废弃，转而被替换成 logging.file.name。
- logging.path：指定日志输出的目标路径。与 logging.file 一起使用时，将只有一项生效。该属性同样在 2.2.x 及以上版本被替换，替换后的属性名为 logging.file.path。
- logging.pattern.console：指定日志在控制台输出的格式。
- logging.patter.file：指定日志在文件输出的格式。

2.4.3 详细配置

尽管无配置或者在 application.yml 文件中进行配置的方式很有用，但它很可能不能满足我们的日常需求。为了应对更复杂的需求，Spring Boot 支持独立配置的方式。当 Classpath 中包含以下几种配置文件时，Spring Boot 将会默认自动加载它们：

- logback-spring.xml
- logback.xml
- logback-spring.groovy
- logback.groovy

其中使用"-spring"形式的配置文件是官方更为推荐的形式。以下是一个 logback-spring.xml 的示例：

```xml
<?xml version="1.0" encoding="UTF-8"?>
<configuration>

    <property name="LOGS" value="./logs" />

    <appender name="Console"
        class="ch.qos.logback.core.ConsoleAppender">
        <layout class="ch.qos.logback.classic.PatternLayout">
            <Pattern>
                %black(%d{ISO8601}) %highlight(%-5level)
[%blue(%t)] %yellow(%C{1.}):
    %msg%n%throwable
```

```xml
            </Pattern>
        </layout>
    </appender>

    <appender name="RollingFile"
        class="ch.qos.logback.core.rolling.RollingFileAppender">
        <file>${LOGS}/spring-boot-logger.log</file>
        <encoder
            class="ch.qos.logback.classic.encoder.PatternLayoutEncoder">
            <Pattern>%d %p %C{1.} [%t] %m%n</Pattern>
        </encoder>

        <rollingPolicy
            class="ch.qos.logback.core.rolling.TimeBasedRollingPolicy">
            <!-- rollover daily and when the file reaches 10 MegaBytes -->
<fileNamePattern>${LOGS}/archived/spring-boot-logger-%d{yyyy-MM-dd}.%i.log
            </fileNamePattern>
            <timeBasedFileNamingAndTriggeringPolicy
                class="ch.qos.logback.core.rolling.SizeAndTimeBasedFNATP">
                <maxFileSize>10MB</maxFileSize>
            </timeBasedFileNamingAndTriggeringPolicy>
        </rollingPolicy>
    </appender>

    <!-- 设置全局日志等级为 info -->
    <root level="info">
        <appender-ref ref="RollingFile" />
        <appender-ref ref="Console" />
    </root>

    <!-- 将 com.example 包下的日志等级设置为 trace -->
    <logger name="com.example" level="trace" additivity="false">
        <appender-ref ref="RollingFile" />
        <appender-ref ref="Console" />
    </logger>

</configuration>
```

2.4.4　Lombok 注解：@Sl4j 和 @Commonslog

在编写日志相关信息时，总是需要借助 LoggerFactory 获取一个 Logger 实例。为了消除这一类模板代码，Lombok 同样提供了两个注解：@Sl4j 以及 @Commonslog。使用 @Sl4j 修改 LogController.java：

```
@RequestMapping("/log")
@Slf4j
@RestController
public class LogController {

    @GetMapping
    public String justShowSomeLog() {
        log.trace("TRACE Message.");
```

```
        log.debug("DEBUG Message.");
        log.info("INFO Message.");
        log.warn("WARN Message.");
        log.error("ERROR Message.");
        return "Bro,go check your console.";
    }
}
```

使用@Commonslog 替代@Sl4j，结果相同。

2.4.5 在 Windows 平台输出彩色日志的 JANSI

分别在 Windows 平台与类 Unix 平台启动 Spring Boot 项目，可以观察到控制台上日志输出形式存在差异。在类 Unix 平台的控制台输出的日志默认是以彩色显示的，而在 Windows 平台输出的日志是以单色调显示的。原因在于类 Unix 平台默认支持 ANSI 色彩，而 Windows 却不能。因此，在 Windows 平台的控制台上需要输出彩色日志的话，需要用到 JANSI 这个库。在 pom.xml 中新增 JANSI 的依赖：

```
<dependency>
    <groupId>org.fusesource.jansi</groupId>
    <artifactId>jansi</artifactId>
    <version>1.18</version>
</dependency>
```

并在 logback-spring.xml 中启用它：

```
<configuration debug="true">
    <appender name="STDOUT" class="ch.qos.logback.core.ConsoleAppender">
        <withJansi>true</withJansi>
        <encoder>
            <pattern>%d{yyyy-MM-dd HH:mm:ss.SSS} [%thread] %highlight(%-5level) %green(%logger{15}) - %msg %n</pattern>
        </encoder>
    </appender>
    <root level="INFO">
        <appender-ref ref="STDOUT" />
    </root>
</configuration>
```

这样一来，在 Windows 平台上运行 Spring Boot 项目，其控制台的日志输出也会以彩色的形式显示出来了。

2.5 过滤器与拦截器

在实际的开发过程中，可能会遇到这样一类需求：统计在线用户、敏感词过滤或者基于 URL 进行访问控制。这些需求有一个共同点——在每个接口被请求时都需要进行该类操作。换而言之，如果编写了对应以上需求的代码，在每一个接口的某处都需要对这些代码进行调用。不使用一些技巧的话，这个开发过程会变得异常烦琐。本节介绍的 Filter（过滤器）与 Interceptor（拦截器）将合理解决这类需求。

2.5.1 过滤器

Filter（过滤器）这一概念来源于"Servlet 规范"，具体的功能实现由 Servlet 容器（即 Spring Boot 内容的 Tomcat）提供。过滤器的主要职责在于对资源的请求与响应的过滤，对从客户端向服务端发送的请求进行过滤，也可以对服务端返回的响应进行处理。Filter 与 Servlet 是有区别的。Filter 虽然可以对请求与响应做出处理，但其本身并不可以产生响应，如图 2.7 所示。

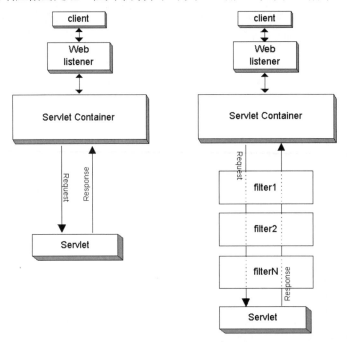

图 2.7 Filter 工作流程示意图

2.5.2 使用过滤器实现访问控制

【示例 2-6】

（1）创建一个需要实施访问控制的控制器。在路径 src/main/java/com/example/myb-log/controller 下创建 SecretController.java：

```
@RequestMapping("/secret")
@RestController
public class SecretController {

    @GetMapping
public String secret() {
    //以下代码可以替换成任意需要被保护的内容
```

```
        return "secret";
    }
}
```

（2）创建用于身份认证的控制器。如果通过了认证，控制器将在 Cookie 中写入作为身份凭证的内容。在 controller 路径下创建 SessionController.java：

```
@RestController
@RequestMapping("/session")
public class SessionController {

    @PostMapping
    public String login(@RequestBody SessionQuery sessionQuery,
HttpServletResponse response) {
        if (authenticate(sessionQuery)) {
            certificate(response);
            return "success";
        }
        //登录失败返回错误
        return "failed";
    }

    private boolean authenticate(SessionQuery sessionQuery) {
        //简单的验证逻辑，仅用作演示
        return Objects.equals(sessionQuery.getUsername(), "admin") &&
            Objects.equals(sessionQuery.getPassword(), "password");
    }

    private void certificate(HttpServletResponse response) {
        //将登录凭证以 Cookie 的形式返回给客户端
        Cookie credential = new Cookie("sessionId", "test-token");
        response.addCookie(credential);
    }

}
```

（3）创建用于检查凭证的过滤器。过滤器通过检查请求中附带的 Cookie 内容，以确认用户的身份。在路径 src/main/java/com/example/myblog/filter 下创建 SessionFilter.java：

```
@Slf4j
@WebFilter(urlPatterns = "/secret/*")
public class SessionFilter implements Filter {
    @Override
    public void doFilter(ServletRequest servletRequest, ServletResponse
servletResponse, FilterChain filterChain) throws IOException, ServletException {
        //读取 Cookie
        Cookie[] cookies = Optional.ofNullable(((HttpServletRequest)
servletRequest).getCookies())
                .orElse(new Cookie[0]);
        boolean unauthorized = true;
        for (Cookie cookie : cookies) {
            if ("sessionId".equals(cookie.getName()) &&
"test-token".equals(cookie.getValue())) {
                //验证 Cookie 中的凭证内容，如果通过验证则继续执行，否则返回 401 错误
```

```
                unauthorized = false;
            }
        }
        if (unauthorized) {
            log.error("UNAUTHORIZED");
            unauthorizedResp(servletResponse);
        }else {
            filterChain.doFilter(servletRequest, servletResponse);
        }
    }

    //向响应中写入 401 错误
    private void unauthorizedResp(ServletResponse response) throws IOException {
        HttpServletResponse httpResponse = (HttpServletResponse) response;
        httpResponse.setStatus(HttpServletResponse.SC_UNAUTHORIZED);
        httpResponse.setHeader("Content-type", "text/html;charset=UTF-8");
        httpResponse.setCharacterEncoding("UTF-8");
        httpResponse.getWriter().write("UNAUTHORIZED");
    }
}
```

（4）启用过滤器。需要在主类中使用@ServletComponentScan 注解，以启用被注解@WebFilter 修饰的过滤器。

```
@SpringBootApplication
@EnableConfigurationProperties({BlogProperties.class,
FileStorageProperties.class})
@ServletComponentScan(basePackages = {"com.example.myblog.filter"})
public class BlogApplication {
    public static void main(String[] args) {
        SpringApplication.run(BlogApplication.class, args);
    }
}
```

（5）分别在请求登录接口前后访问"受保护"的路径，以确认访问控制是否生效。

2.5.3 拦截器

Interceptor（拦截器）这一功能由 Spring 提供。Interceptor 与 Filter 类似，操作粒度更小，但整体功能不如 Filter 强大。Interceptor 支持自定义预处理（preHandle）可以在此过程中决定是否禁止程序继续进行，自定义后续处理（postHandle）。其处理流程如图 2.8 所示。

使用拦截器的前提是需要实现 HandlerInterceptor 接口。该接口包含三种主要方法：

- preHandle()：在执行实际的处理程序之前调用，但尚未生成视图。
- postHandle()：处理程序执行后调用。
- afterCompletion()：在请求已经响应并且视图生成完毕之后调用。

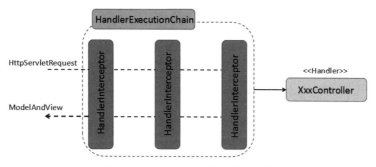

图 2.8　Interceptor 工作流程示意图

2.5.4　使用拦截器记录请求参数

首先创建一个继承 HandlerInterceptorAdapter 的拦截器类，使用日志打印请求中的参数。在路径 src/main/java/com/example/myblog/interceptor 下创建 LogRequestInterceptor.java：

【示例 2-7】

```java
@Slf4j
@Component
public class LogRequestInterceptor extends HandlerInterceptorAdapter {

    @Override
    public boolean preHandle(HttpServletRequest request, HttpServletResponse response, Object handler) throws Exception {
        log.info(String.format("[preHandle][%s][%s]%s%s", request,
request.getMethod(), request
    .getRequestURI(), getParameters(request)));
        return true;
    }

    @Override
    public void postHandle(HttpServletRequest request, HttpServletResponse response, Object handler, ModelAndView modelAndView) throws Exception {
        log.info("[postHandle]");
    }

    @Override
    public void afterCompletion(HttpServletRequest request,
HttpServletResponse response, Object handler, Exception ex) throws Exception {
        if (ex != null) {
            ex.printStackTrace();
        }
        log.info(String.format("[afterCompletion][%s][exception:%s]",
request, ex));
    }

    //提取请求中的参数
    private String getParameters(HttpServletRequest request) {
        StringBuilder parameterBuilder = new StringBuilder();
        Enumeration<String> names = request.getParameterNames();
        if (names != null) {
```

```
                parameterBuilder.append("?");
            while (names.hasMoreElements()) {
                if (parameterBuilder.length() > 1) {
                    parameterBuilder.append("&");
                }
                String pointer = names.nextElement();
                parameterBuilder.append(pointer).append("=").
append(request.getParameter(pointer));
            }
        }
        return parameterBuilder.toString();
    }
}
```

然后将拦截器配置到 Spring 上下文。在路径 src/main/java/com/example/myblog/config 下创建 InterceptorConfig.java：

```
@Configuration
@RequiredArgsConstructor
public class InterceptorConfig implements WebMvcConfigurer {

    private final LogRequestInterceptor logRequestInterceptor;

    @Override
    public void addInterceptors(InterceptorRegistry registry) {
        registry.addInterceptor(logRequestInterceptor).
addPathPatterns("/**");
    }
}
```

最后，任意访问一个已发布过的接口。如果拦截器已生效，将会看到上文编写的日志打印到控制台当中。

2.6 Spring Boot 事件

当业务变得繁杂，模块与模块之间的耦合变得愈发严重的时候，很自然的会想要去对该部分代码进行解耦。"事件驱动"这一架构在解耦方面是一把好手。Java 中已经对事件驱动提供了支持，这种架构在 Swing 的 GUI 编程中比较常见。Spring Boot 在此基础上扩展出了自己的事件机制。

2.6.1 事件驱动模型

事件驱动模型由三个核心部分组成，如图 2.9 所示。

图 2.9 事件驱动模型

- EventSource（事件源）：事件发生的场所（对象）。
- Event（事件）：对事件信息的封装。将其视为一种通知会更易理解一些。
- EventListener（事件监听器）：负责监听事件（事件通知）并对其做出反应的组件。

当业务领域内有状态发生变化时，可以通过发送事件通知的方式通知其他模块。这种模式的特点在于事件源并不关心外部系统的实现，也并不期待通知的发送会带来何种的结果。换而言之，事件源相关的代码与事件监听器相关的代码没有"直接调用"的这种关系。

没有直接调用，这种方式换来了非常容易实现的"低耦合"。例如，一种爆款商品正在热卖中，这个售卖的业务中有"营销"、"售后"以及"库存管理"各种不同的部门参与其中。商品的状态由"销售中"转变为"售罄"，这一事件被发送出去并被各部门的"监听器"监听到后，便可以开始执行各个部门自有的业务逻辑。其中，这个业务若发生变化，例如有部门在业务逻辑中新增或者删除，只需修改对应的监听器即可。

不过，这样的松耦合也是一把双刃剑。因为没有直接调用，代码中缺少对流程的显式描述。如果业务流程变得复杂，整个过程会变得难以调试与修改，最终在系统中留下隐患。

2.6.2 内置事件

Spring 内置了五种标准上下文事件以及四种 ApplicationContextEvent，以便于开发人员进入应用程序以及上下文的生命周期并执行一些自定义操作。当然，即便我们很少手动使用这些事件，该框架内也会大量使用它们：

- ContextRefreshedEvent：上下文更新事件。该事件会在 ApplicationContext 被初始化或者更新时发布。也可以在调用 ConfigurableApplicationContext 接口中的 refresh() 方法时被触发。
- ContextStartedEvent：上下文开始事件。当调用 ConfigurableApplicationContext 的 Start() 方法开始或重启容器时触发该事件。
- ContextStoppedEvent：上下文停止事件。当调用 ConfigurableApplicationContext 的 Stop() 方法停止容器时触发该事件。
- ContextClosedEvent：上下文关闭事件。当 ApplicationContext 被关闭时触发该事件。容器被关闭时，其管理的所有单例 Bean 都被销毁。
- RequestHandledEvent：请求处理事件。在 Web 应用中，当一个 http 请求结束时触发该事件。
- ApplicationStartedEvent：应用启动事件 Spring Boot 启动开始时触发该事件。
- ApplicationEnvironmentPreparedEvent：应用环境就绪事件。该事件在环境（Environment）准备就绪而上下文还没准备就绪的情况下触发。
- ApplicationPreparedEvent：应用就绪事件。该事件在上下文准备就绪而 Bean 还没加载完成时触发。
- ApplicationFailedEvent：应用异常事件。该事件在 Spring Boot 启动异常时触发。

2.6.3 监听内置事件

监听内置事件仅需要创建事件对应的监听器并注册即可，以监听 ApplicationPreparedEvent 为例。

(1) 在路径 src/main/java/com/example/myblog/event 下创建 CustomApplicationPreparedEventListener.java：

```java
@Slf4j
public class CustomApplicationPreparedEventListener implements
ApplicationListener<ApplicationPreparedEvent> {

    @Override
    public void onApplicationEvent(ApplicationPreparedEvent
applicationPreparedEvent) {
        log.info("Gotya! ApplicationPreparedEvent.");
    }

}
```

(2) 注册监听器。该过程需要对 BlogApplication 进行改动：

```java
@SpringBootApplication
@EnableConfigurationProperties({BlogProperties.class,
FileStorageProperties.class})
@ServletComponentScan(basePackages = {"com.example.myblog.filter"})
public class BlogApplication {

    public static void main(String[] args) {
        SpringApplication application = new
SpringApplication(BlogApplication.class);
        application.addListeners(new
CustomApplicationPreparedEventListener());
        application.run(args);
    }

}
```

(3) 修改完成之后重启应用，即可看到对应日志成功输出至控制台。

2.6.4 自定义事件

使用 Spring Boot 事件进行代码解耦，离不开自定义事件这一功能。下面来介绍自定义事件的实现过程：

(1) 创建自定义事件类。在路径 src/main/java/com/example/myblog/event 下创建 MessageEvent.java：

```java
@Getter
public class MessageEvent extends ApplicationEvent {

    private final String message;

    public MessageEvent(Object source, String message) {
        super(source);
        this.message = message;
```

（2）创建事件源。在 event 路径创建 MessageEventPublisher.java：

```
@Slf4j
@Component
@RequiredArgsConstructor
public class MessageEventPublisher {

    private final ApplicationEventPublisher applicationEventPublisher;

    public void publishAnEvent(String message) {
        log.info("Publishing an event.Message:" + message);
        //发送事件通知
        applicationEventPublisher.publishEvent(new MessageEvent(this, message));
    }

}
```

（3）创建事件监听器。事件监听器的创建有两种方式，使用实现 ApplicationListener<E extends ApplicationEvent>接口的方式创建 MessageEventListener.java：

```
@Slf4j
@Component
public class MessageEventListener implements ApplicationListener<MessageEvent> {
    @Override
    public void onApplicationEvent(MessageEvent messageEvent) {
        //处理接收到事件通知后的业务逻辑
        log.info("Some business……Message:" + messageEvent.getMessage());
    }
}
```

使用@EventListener 这一注解的方式创建 AnotherMessageEventListener.java：

```
@Slf4j
@Component
public class AnotherMessageEventListener {

    @EventListener
    public void onApplicationEvent(MessageEvent messageEvent) {
        //处理接收到事件通知后的业务逻辑
        log.info("Other business……Message:" + messageEvent.getMessage());
    }
}
```

（4）使用测试代码测试自定义事件监听相关逻辑，在路径 src/test/java/co-m/example/myblog 下创建 EventTests.java：

```
@SpringBootTest(classes = {BlogApplication.class},
webEnvironment = SpringBootTest.WebEnvironment.RANDOM_PORT)
public class EventTests {

    @Autowired
```

```
    MessageEventPublisher publisher;

    @Test
    public void publishAnEvent_thenCheckConsole() {
        publisher.publishAnEvent("Bada bing,bada boom......");
    }
}
```

在测试代码启动之后,可以在控制台观察到事件相关的日志被打印到控制台。这意味着相关逻辑已成功执行。

2.6.5 异步事件

在默认情况下,事件的发布与监听是同步执行的。当要用到异步事件时,需要进行额外的配置。具体方式在于创建 ApplicationEventMulticaster 的 JavaBean。在路径 src/main/java/com/example/myblog/config 下创建 AsynchronousEventsConfig.java:

```
@Configuration
public class AsynchronousEventsConfig {
    @Bean(name = "applicationEventMulticaster")
    public ApplicationEventMulticaster simpleApplicationEventMulticaster() {
        SimpleApplicationEventMulticaster eventMulticaster = new SimpleApplicationEventMulticaster();
        eventMulticaster.setTaskExecutor(new SimpleAsyncTaskExecutor());
        return eventMulticaster;
    }
}
```

这样一来监听器将会在单独的进程中执行,如图 2.10 所示。

```
INFO 26596 --- [           main] c.e.myblog.event.MessageEventPublisher   : Publishing an event.Message:Bada bing,bada boom.
INFO 26596 --- [TaskExecutor-38] c.e.m.event.AnotherMessageEventListener  : Other business......Message:Bada bing,bada boom.
INFO 26596 --- [TaskExecutor-39] c.e.myblog.event.MessageEventListener    : Some business......Message:Bada bing,bada boom.
```

图 2.10 监听器在不同监听器中执行

第 3 章

创建 RESTful Web 服务

随着移动互联网的发展，Web 开发技术的更迭，前后端分离的软件设计架构在 Web 应用开发中被广泛应用。为了能构建出更为可靠易用的服务端程序，HTTP 规范制定人之一 Dr. Roy Thomas Fielding 设计了一套 REST 规范（Representational State Transfer）。这套规范一言以蔽之，就是 URL 用以定位资源，HTTP 动词用以描述操作。后端接口的设计，就这样被安排得明明白白的。作为 Web 开发界的"扛把子"，Spring Boot 自然也提供了 REST 相关的一系列支持。接下来请跟着笔者一起来创建一个 RESTful Web 服务吧。

本章主要涉及的知识点有：

- HTTP 动词以及相关接口设计的概念
- 如何进行前后端的数据交互
- 在后端接口进行参数验证以及错误处理
- 使用 Swagger 生成方便测试与对接的接口文档

3.1 HTTP 动词

使用 REST 规范来构建后端应用的关键点，在于需要使用 URL 和 HTTP 动词来描述"调用方"与"资源"的交互。首先来认识一下常见的 HTTP 动词。

- GET：从服务端取出资源。
- POST：在服务端新建资源。
- PUT：在服务端更新资源（指客户端提供更改后完整的资源）。
- PATCH：在服务端更新资源（指客户端提供改变的属性）。
- DELETE：从服务端删除资源。

使用以上 HTTP 动词，结合合适的 URL 定义，基本上可以覆盖对于资源的各种操作。例如：

- GET /vehicle/list：获取 Vehicle 记录列表。
- GET / vehicle /{id}：根据 id 获取某 Vehicle 的信息。
- POST / vehicle：新建 Vehicle 记录。
- PUT / vehicle /{id}：整个地更新/替换某 Vehicle 的信息。
- PATCH / vehicle /{id}：修改 Vehicle 记录的某片段。
- DELTE / vehicle /{id}：删除对应的 Vehicle 记录。

现在我们可以依据这个 API 设计着手构建一个基础的 RESTful 后端应用。

3.1.1　构建一个基础的 RESTful Web 服务

在编写代码之前，需要做一下准备工作。通过 Spring Initializr（https://start.spring.io/）创建一个初始化工程，在这个页面选择项目将用到的依赖项，比如当前项目将会使用到 Spring Web、H2 Database 和 Spring Data JPA 这三项。选择完毕后在页面下方单击 Generate 按钮即可获取一个空的初始化工程，如图 3.1 所示。

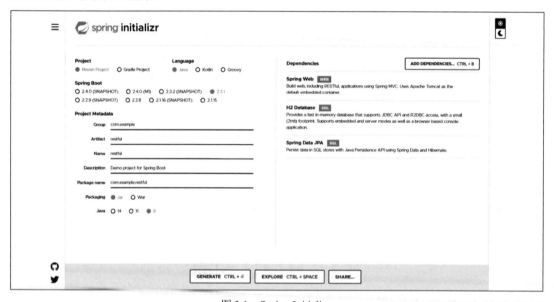

图 3.1　Spring Initializr

【示例 3-1】

打开初始化工程，开始着手编写代码。上文提到的接口被设计用于管理一些 Vehicle 记录，所以首先需要定义代表这些信息的实体类 Vehicle。

```
package com.example.restful;

import javax.persistence.*;

@Entity
@Table(name = "t_vehicle")
```

```java
public class Vehicle {

    @Id
    @GeneratedValue(strategy = GenerationType.IDENTITY)
    private Long id;

    private String name;

    private String description;

    public Long getId() {
        return id;
    }

    public void setId(Long id) {
        this.id = id;
    }

    public String getName() {
        return name;
    }

    public void setName(String name) {
        this.name = name;
    }

    public String getDescription() {
        return description;
    }

    public void setDescription(String description) {
        this.description = description;
    }

    @Override
    public String toString() {
        return "Vehicle{" +
            "id=" + id +
            ", name='" + name + '\'' +
            ", description='" + description + '\'' +
            '}';
    }
}
```

定义好实体类之后，需要定义一个 JpaRepository 接口 VehicleRepository 用于操作数据库。Spring Data JPA 的具体使用方法会在后面的章节详细介绍。这里了解声明对应的 JpaRepository 可以做基础的 CURD 操作即可。

```java
package com.example.restful;

import org.springframework.data.jpa.repository.JpaRepository;

public interface VehicleRepository extends JpaRepository<Vehicle, Long> {
}
```

声明好了实体类以及对应的 JpaRepository 类，我们可以自由地在代码中对数据库进行操作了。现在需要将这些功能根据之前的接口设计实现成 REST Web 服务。最后再实现一个 RestController。

```java
package com.example.restful;

import org.springframework.beans.BeanUtils;
import org.springframework.beans.BeanWrapper;
import org.springframework.beans.BeanWrapperImpl;
import org.springframework.http.HttpStatus;
import org.springframework.http.ResponseEntity;
import org.springframework.web.bind.annotation.*;

import java.beans.FeatureDescriptor;
import java.util.List;
import java.util.Objects;
import java.util.Optional;
import java.util.stream.Collectors;
import java.util.stream.Stream;

@RestController
@RequestMapping("/vehicle")
public class VehicleController {

    private final VehicleRepository vehicleRepository;

    public VehicleController(VehicleRepository vehicleRepository) {
        this.vehicleRepository = vehicleRepository;
    }

    @GetMapping("/list")
    public ResponseEntity<List<Vehicle>> vehicleList() {
        //列出所有 Vehicle 记录
        return new ResponseEntity<>(vehicleRepository.findAll(), HttpStatus.OK);
    }

    @GetMapping("/{id}")
    public ResponseEntity<Vehicle> selectOne(@PathVariable Long id) {
        //根据 Id 查出一条 Vehicle 记录
        return vehicleRepository.findById(id).map(v -> new ResponseEntity<>(v, HttpStatus.OK))
                .orElse(new ResponseEntity<>(null, HttpStatus.BAD_REQUEST));
    }

    @PostMapping("/")
    public ResponseEntity<Vehicle> createOne(@RequestBody Vehicle vehicle) {
        //创建一条 Vehicle 记录
        return new ResponseEntity<>(vehicleRepository.save(vehicle), HttpStatus.OK);
    }

    @PutMapping("/")
    public ResponseEntity<Vehicle> replaceOne(@RequestBody Vehicle vehicle) {
        //替换一条 Vehicle 记录
        Optional<Vehicle> oldOne =
```

```java
vehicleRepository.findById(vehicle.getId());
        if (!oldOne.isPresent()) {
            new ResponseEntity<>(null, HttpStatus.BAD_REQUEST);
        }
        return new ResponseEntity<>(vehicleRepository.save(vehicle),
HttpStatus.OK);
    }

    @PatchMapping("/")
    public ResponseEntity<Vehicle> modifyOne(@RequestBody Vehicle vehicle) {
        //修改 Vehicle 记录
        Optional<Vehicle> findById =
vehicleRepository.findById(vehicle.getId());
        Vehicle oldOne;
        if (!findById.isPresent()) {
            return new ResponseEntity<>(null, HttpStatus.BAD_REQUEST);
        } else {
            oldOne = findById.get();
        }
        Vehicle newOne = new Vehicle();
        List<String> nullProperties = getNullProperties(oldOne);
        BeanUtils.copyProperties(newOne, oldOne, nullProperties.toArray(new
String[0]));
        return new ResponseEntity<>(vehicleRepository.save(newOne),
HttpStatus.OK);
    }

    @DeleteMapping("/{id}")
    public ResponseEntity<Vehicle> deleteOne(@PathVariable Long id) {
        //根据 Id 删除一条记录
        vehicleRepository.deleteById(id);
        return new ResponseEntity<>(null, HttpStatus.OK);
    }

    private List<String> getNullProperties(Object source) {
        //获取空属性对应的属性名
        final BeanWrapper wrappedSource = new BeanWrapperImpl(source);
        return Stream.of(wrappedSource.getPropertyDescriptors())
                .map(FeatureDescriptor::getName)
                .filter(propertyName ->
Objects.isNull(wrappedSource.getPropertyValue(propertyName)))
                .collect(Collectors.toList());
    }
}
```

每一个 RestController 类都需要被 @RestController 这个注解修饰。这个注解等同于 @ResponseBody+@Controller。最后呈现出的效果就是，该控制器下暴露出来的所有接口都是会自动序列化成 JSON 格式，并放进 HttpResponse 中。@GetMapping@PostMapping 这些注解即对应 HTTP 动词。

3.1.2　是 GetMapping 吗？是 RequestMapping

从以上的例子可以观察到，通过@GetMapping、@PostMapping、@PutMapping、@PatchMapping以及@DeleteMapping 可以定义一个接口的 URL 并限定它能接受的 HTTP 方法。那它还有别的功能吗？让我们打开其中一个注解@GetMapping 的源码来一窥究竟。（使用 IDEA 的读者，可以通过 Ctrl+鼠标左键选中@GetMapping 的方式打开 GetMapping 反编译后的代码。单击"Download Source"按钮即可下载并查看源码。）

【示例 3-2】

```java
package org.springframework.web.bind.annotation;

import java.lang.annotation.Documented;
import java.lang.annotation.ElementType;
import java.lang.annotation.Retention;
import java.lang.annotation.RetentionPolicy;
import java.lang.annotation.Target;

import org.springframework.core.annotation.AliasFor;

/**
 * Annotation for mapping HTTP {@code GET} requests onto specific handler
 * methods.
 *
 * <p>Specifically, {@code @GetMapping} is a <em>composed annotation</em> that
 * acts as a shortcut for {@code @RequestMapping(method = RequestMethod.GET)}.
 *
 * @author Sam Brannen
 * @since 4.3
 * @see PostMapping
 * @see PutMapping
 * @see DeleteMapping
 * @see PatchMapping
 * @see RequestMapping
 */
@Target(ElementType.METHOD)
@Retention(RetentionPolicy.RUNTIME)
@Documented
@RequestMapping(method = RequestMethod.GET)
public @interface GetMapping {

    /**
     * Alias for {@link RequestMapping#name}.
     */
    @AliasFor(annotation = RequestMapping.class)
    String name() default "";

    /**
     * Alias for {@link RequestMapping#value}.
     */
    @AliasFor(annotation = RequestMapping.class)
```

```
    String[] value() default {};

    /**
     * Alias for {@link RequestMapping#path}.
     */
    @AliasFor(annotation = RequestMapping.class)
    String[] path() default {};

    /**
     * Alias for {@link RequestMapping#params}.
     */
    @AliasFor(annotation = RequestMapping.class)
    String[] params() default {};

    /**
     * Alias for {@link RequestMapping#headers}.
     */
    @AliasFor(annotation = RequestMapping.class)
    String[] headers() default {};

    /**
     * Alias for {@link RequestMapping#consumes}.
     * @since 4.3.5
     */
    @AliasFor(annotation = RequestMapping.class)
    String[] consumes() default {};

    /**
     * Alias for {@link RequestMapping#produces}.
     */
    @AliasFor(annotation = RequestMapping.class)
    String[] produces() default {};

}
```

打开源码，可以看到源码中的注释。其实这些"Mapping"本质上就是 RequestMapping。可以观察到这个注解包含有非常多的属性，这里挑选几个相对实用的属性来简单介绍一下。

- path：指定接口的 URL 访问路径。
- params：指定请求中必须包含的参数。
- headers：指定请求中必须包含的请求头。
- consumes：指定请求的内容类型（Content-Type）。
- produces：指定响应的内容类型（Content-Type）。

现在请来设想一个场景。出现了一个新的需求，需要在查询到一条 Vehicle 记录之后，将这条信息复制并再次插入到数据库当中。单纯依靠 URL 以及 HTTP 动词，好像并不能很好地完成这项任务。不要担心，这时候 params 可以派上用场了。

```
    @PostMapping(path = "/{id}", params = "method")
    public ResponseEntity<Vehicle> dependOnMethod(@PathVariable Long id, String method) {
        switch (method) {
            case "select": {
```

```java
        return selectOne(id);
    }
    case "delete": {
        return deleteOne(id);
    }
    case "duplicate": {
        return duplicateOne(id);
    }
    default: {
        return new ResponseEntity<>(null, HttpStatus.BAD_REQUEST);
    }
    }
}

private ResponseEntity<Vehicle> duplicateOne(Long id) {
    //复制一条记录并写入到数据库当中
    Optional<Vehicle> findById = vehicleRepository.findById(id);
    if (!findById.isPresent()) {
        return new ResponseEntity<>(null, HttpStatus.BAD_REQUEST);
    } else {
        Vehicle oldOne = findById.get();
        Vehicle newOne = new Vehicle();
        newOne.setName(oldOne.getName());
        newOne.setDescription(oldOne.getDescription());
        return new ResponseEntity<>(vehicleRepository.save(newOne), HttpStatus.OK);
    }
}
```

在这里扩展了一个 method 参数，通过传入不同的 method 参数实现不同的需求。当参数为"duplicate"的时候，复制对应的 Vehicle 实体中的内容，并插入到数据库中。

这样一来，一个基础的 REST 服务就构建出来了。程序成功运行之后，可以通过 curl 或者 Postman 这类工具简单测试与检验。在 3.4 节将介绍另外一个非常酷的方式来与接口交互。当然了，一个稳定可靠的 REST 服务只有这些是远远不够的。不用着急，我们在后续的章节会慢慢完善这个程序。

3.2 请求与响应

作为信息系统的一员，Web 服务的数据交互无论在开发或使用中，都是备受关注的方面。其中与用户（客户端）的数据交互大多需要依赖 HTTP 协议中的请求与响应。本节将对此展开介绍。

3.2.1 HTTP 报文

请求（Request）与响应（Response）属于 HTTP 报文的两种形式，由客户端传递至服务端就称为请求，由服务端返回客户端则称为响应。同为 HTTP 报文，它们都有如图 3.2 所示的结构。

- 起始行：用于描述请求或者响应的状态。
- Header：HTTP 头信息。
- 空行：由 CRLF 字符组成的空行，用于分隔 HTTP 头与 HTTP 报文体。
- Body：报文体，用于搭载请求或响应中的实体。

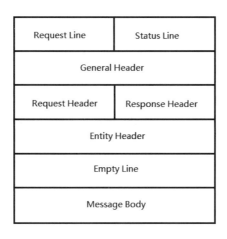

图 3.2　HTTP 报文结构

3.2.2　简单请求与@RequestParam

当请求内容相对简单时，可以考虑使用普通参数用于提取请求中的内容。代码示例如下：

【示例 3-3】

```
@GetMapping("/no-param")
public String noParam() {
    //无参
    return "No parameter.";
}

@GetMapping(path = "/single-param")
public String singleParam(String param) {
    //单个可选参数
    return "The parameter is :" + param;
}

@GetMapping("/single-param-with-annotation")
public String singleParamWithAnnotation(@RequestParam("parameter") String param) {
    //单个必传参数
    return "The parameter is :" + param;
}

@PostMapping("/single-param-with-default-value")
public String singleParamWithDefaultValue(@RequestParam(defaultValue = "default") String param) {
    //单个必传参数
    return "The parameter is :" + param;
```

```java
}

@PostMapping("/few-params-with-annotation")
public String fewParamsWithAnnotation(@RequestParam String paramA,
@RequestParam(required = false) Integer paramB) {
    //paramA 必传, paramB 可选
    return String.format("paramA is :%s,paramB is :%s", paramA, paramB);
}
```

@RequestParam 注解在该场景下可以很好地协助开发人员。虽然该注解为可选项，不使用也不会影响参数的映射，不过它提供了三个非常实用的属性：name（Web 参数名）、required（是否必传）以及 defaultValue（默认值）。使用这些属性，可以使接口定义这一开发环节变得灵活方便。

> **注　意**
>
> @RequestParam 的 required 属性默认为 true。这意味着如果未接收到对应参数或对应参数值为 null 的话，接口将会报错。

一个引发报错的请求示例：

```
POST http://localhost:8080/api/simple/few-params-with-annotation HTTP/1.1
User-Agent: PostmanRuntime/7.26.3
Accept: */*
Postman-Token: da5ca131-f063-444a-8195-f1433a31974e
Host: localhost:8080
Accept-Encoding: gzip, deflate, br
Connection: keep-alive
Content-Type: application/x-www-form-urlencoded
Content-Length: 9

paramB=13
```

响应结果：

```
HTTP/1.1 400
Content-Type: application/json
Transfer-Encoding: chunked
Date: Wed, 19 Aug 2020 15:23:19 GMT
Connection: close

14e3
{"timestamp":"2020-08-19T15:23:19.337+00:00","status":400,"error":"Bad Request"…
```

不过有一种情况是例外——使用 Optional 作为入参。修改过后的代码：

```java
@PostMapping("/few-params-with-annotation")
public String fewParamsWithAnnotation(@RequestParam Optional<String> paramA,
@RequestParam(required = false) Integer paramB) {
    //paramA 必传, paramB 可选
    return String.format("paramA is :%s,paramB is :%s", paramA.orElse(null), paramB);
}
```

使用相同请求，返回不同的响应结果：

```
HTTP/1.1 200
Content-Type: text/plain;charset=UTF-8
Content-Length: 29
Date: Wed, 19 Aug 2020 15:32:32 GMT
Keep-Alive: timeout=60
Connection: keep-alive

paramA is :null,paramB is :13
```

不过这并不是一个好的实现,此处仅用作简要说明。

3.2.3 使用@PathVariable 获取 URL 中的参数

根据 RESTful 风格设计出来的 API 接口,"URL 中包含查询所需的元素"这种情况也是常见的。例如"/user/{username}",该 URL 对应的接口根据占位符中内容的不同,展示不同的用户信息。相比于另一种形式"/user?username=...",这种处理参数的方式会让 URL 的设计更加简洁直观。要实现这种风格的接口就需要借助@PathVariable 注解。

事实上,在之前的 UserController 中已经对该注解有所接触。接下来将继续之前的示例,以展示@PathVariable 的多种使用方法。

【示例 3-4】

```
@GetMapping("/{loqin}")
public User findOne(@PathVariable String login) {
    //单个参数绑定单个入参
    User result = userRepository.findByLogin(login);
    if (result == null) {
        throw new ResponseStatusException(HttpStatus.NOT_FOUND, "This user does not exit");
    }
    return result;
}

@GetMapping(value = "/{firstName}/{lastName}", params = MULTI)
public User findOne(@PathVariable String firstName, @PathVariable String lastName) {
    //多个参数绑定多个入参
    User result = userRepository.findByFirstNameAndLastName(firstName, lastName);
    if (result == null) {
        throw new ResponseStatusException(HttpStatus.NOT_FOUND, "This user does not exit");
    }
    return result;
}

@GetMapping(value = "/{firstName}/{lastName}", params = IN_MAP)
public User findOne(@PathVariable Map<String, String> paramMap) {
    //多个参数绑定单个 Map
    User result =
userRepository.findByFirstNameAndLastName(paramMap.get("firstName"),
paramMap.get("lastName"));
```

```
        if (result == null) {
            throw new ResponseStatusException(HttpStatus.NOT_FOUND, "This user does
not exit");
        }
        return result;
    }
```

> **注 意**
>
> 在两个多参数的演示代码中,@GetMapping 中的 params 属性并非必传项,仅在路径重复时作区分用。

除了以上展示的@PathVariable 使用方式之外,该注解结合正则表达式还可以实现参数过滤的功能:

```
@GetMapping("/{login:[\\D]+}")
public User findOne(@PathVariable String login) {
    //login 只匹配内容为非数字的参数
    //单个参数绑定单个入参
    User result = userRepository.findByLogin(login);
    if (result == null) {
        throw new ResponseStatusException(HttpStatus.NOT_FOUND, "This user does
not exit");
    }
    return result;
}
```

3.2.4　借助@RequestHeader 读取请求头

在介绍请求体与传入参数的绑定以及 URL 中参数的获取之后,本小节将介绍如何借助@Req-uestHeader 实现对请求头内容的访问。

【示例 3-5】

(1) 读取单个头属性

如果要读取某个特定的属性,可以使用对应头属性名配置@RequestHeader 的方式:

```
@GetMapping("/greeting")
public String greeting(@RequestHeader("accept-language") String language) {
    //根据 Header 中的不同属性返回不同问候语
    switch (language) {
        case "zh":
            return "你好";
        case "jp":
            return "こんにちは";
        case "en":
        default:
            return "Hello";
    }
}
```

（2）将所有头属性绑定至一个 Map 实例

如果不通过声明获取特定的内容，@RequestHeader 默认获取所有头属性。具体实现方式有三种，其中通过 Map 接收参数最为常见：

```
@GetMapping("/header-map")
public String headerMap(@RequestHeader Map<String, String> headers) {
    //返回一个由所有头属性拼接而成的字符串
    return headers
            .entrySet()
            .stream()
            .map(entry -> String.format("key=%s,value=%s", entry.getKey(), entry.getValue()))
            .collect(Collectors.joining("\r\n"));
}
```

（3）将所有头属性绑定至一个 MultiValueMap 实例

当一个属性拥有多个值时，可以考虑使用 MultiValueMap 实例来接收所有头属性：

```
@GetMapping("/multi-value-map")
public String headerMap(@RequestHeader MultiValueMap<String, String> headers) {
    //返回所有头属性拼接而成的字符串，使用"|"分隔有多个值的属性
    return headers
            .entrySet()
            .stream()
            .map(entry -> String.format("key=%s,value=%s", entry.getKey(), String.join("|", entry.getValue())))
            .collect(Collectors.joining("\r\n"));
}
```

（4）将所有头属性绑定至一个 HttpHeaders 实例

除了以上两种选择之外，还可以通过 HttpHeaders 实例的形式获取所有头属性：

```
@GetMapping("/http-headers")
public String httpHeaders(@RequestHeader HttpHeaders httpHeaders) {
    //获取"Accept-Encoding"属性，以逗号分隔多个值
    return String.join(",",
            Optional.ofNullable(httpHeaders.get("Accept-Encoding"))
                    .orElse(new ArrayList<>()));
}
```

3.2.5 @RequestBody 与 @ResponseBody

这一对注解分别代表着将请求体的内容反序列化为对应类实例（@RequestBody）以及将返回实例序列化为 JSON 字符串（@ResponseBody），在 Restful 服务的开发过程中非常常见。比如 HtmlController.java 中提交文章相关的功能就使用到了这两个注解：

【示例 3-6】

```
@PostMapping("/article")
@ResponseBody
public Article submitArticle(@RequestBody SubmitArticleQuery query) {
```

```java
    //用于接收 POST 请求
    User author = userRepository.findByLogin(query.getAuthor());
    if (author == null) {
        //返回 400 错误码
        throw new ResponseStatusException(HttpStatus.BAD_REQUEST, "This user does not exist");
    }
    Article toSave = new Article().setAuthor(author)
            .setTitle(query.getTitle())
            .setHeadline(query.getHeadline())
            .setContent(query.getContent())
            .setSlug(CommonUtil.toSlug(query.getTitle()));
    articleRepository.save(toSave);
    return toSave;
}
```

返回值的默认 Content-Type 为 application/json。如果需要将其调整为 application/xml，需要引入额外的依赖：

```xml
<dependency>
    <groupId>com.fasterxml.jackson.dataformat</groupId>
    <artifactId>jackson-dataformat-xml</artifactId>
</dependency>
```

微调对应的方法：

```java
@PostMapping(value = "/article", headers = "Accept=application/xml",
produces = MediaType.APPLICATION_XML_VALUE)
@ResponseBody
public Article submitArticleAndGetXml(@RequestBody SubmitArticleQuery query)
{
    //用于接收 POST 请求
    User author = userRepository.findByLogin(query.getAuthor());
    if (author == null) {
        //返回 400 错误码
        throw new ResponseStatusException(HttpStatus.BAD_REQUEST, "This user does not exist");
    }
    Article toSave = new Article().setAuthor(author)
            .setTitle(query.getTitle())
            .setHeadline(query.getHeadline())
            .setContent(query.getContent())
            .setSlug(CommonUtil.toSlug(query.getTitle()));
    articleRepository.save(toSave);
    return toSave;
}
```

3.2.6　使用 ResponseEntity 处理 HTTP 响应

一个 Web 服务的返回值，大多数情况下所关注的仅仅是响应体这一部分，因此响应头以及状态码都是默认状态。如果要对默认的响应头以及状态码进行修改，就需要用到 ResponseEntity 作为返回值去实现。

(1)基础使用方式:

```
@GetMapping("/greeting")
public ResponseEntity<String> greeting() {
    return new ResponseEntity<>("Hello there.", HttpStatus.OK);
}
```

(2)添加自定义的 HTTP header:

```
@GetMapping("/custom-header")
public ResponseEntity<String> customHeader() {
    HttpHeaders headers = new HttpHeaders();
    headers.add("Custom-Header", "customHeader");
    return new ResponseEntity<>("Hello there.", headers, HttpStatus.OK);
}
```

(3)返回不同的状态码:

```
@GetMapping("/next-birth-day/{year}/{month}/{day}")
public ResponseEntity<Long> nextBirthday(@PathVariable int year,@PathVariable int month,@PathVariable int day) {
    LocalDate birthDate = LocalDate.of(year, month, day);
    if (birthDate.isAfter(LocalDate.now())) {
        return new ResponseEntity<>(null, HttpStatus.BAD_REQUEST);
    }
    LocalDate nextBirthDay = LocalDate.of(LocalDate.now().getYear(), birthDate.getMonth(), birthDate.getDayOfMonth());
    if (nextBirthDay.isBefore(LocalDate.now())) {
        nextBirthDay = nextBirthDay.plusYears(1);
    }
    return new ResponseEntity<>(DAYS.between(LocalDate.now(), nextBirthDay),HttpStatus.OK);
}
```

3.3 参数验证

在构建任何程序的过程中,参数验证这一步骤都是难以避免的。比较传统的解决方式通常是在函数或者方法中编写验证相关的业务逻辑。为了将验证从业务代码中抽离,Spring 提供了一种方式——Spring Validation。本节将介绍如何借助 Spring Validation,使参数验证变得简洁通用。

3.3.1 基础验证 Bean Validation

Bean Validation 是 Spring Validation 的基础一环,是由 JCP(Java Community Process)定义的一个标准化的 JavaBean 验证 API。这个 API 提供一组注解,用以标注对应元素的验证形式,但并未提供具体实现。对应功能需要依赖相应的框架工具来实现。其提供的注解如下所示:

- @Null:被标注的元素必须为 Null。
- @NotNull:被标注的元素必须不为 Null。

- @AssertTrue：被标注的元素必须为 True。
- @AssertFalse：被标注的元素必须为 False。
- @Min(value)：被标注的元素必须是一个数字，其值必须大于等于指定的最小值。
- @Max(value)：被标注的元素必须是一个数字，其值必须小于等于指定的最大值。
- @DecimalMin(value)：被标注的元素必须是一个数字，其值必须大于等于指定的最小值。
- @DecimalMax(value)：被标注的元素必须是一个数字，其值必须小于等于指定的最大值。
- @Size(max, min)：被标注的元素的大小必须在指定的范围内。
- @Digits (integer, fraction)：被标注的元素必须是一个数字，其值必须在可接受的范围内。
- @Past：被标注的元素必须是一个过去的日期。
- @Future：被标注的元素必须是一个将来的日期。
- @Pattern(value)：被标注的元素必须符合指定的正则表达式。

3.3.2 高级验证 Spring Validation

Spring Validation 包含 Bean Validation 的实现——Hibernate Validation，并且提供了更多与 Spring 相关的功能。

【示例 3-7】

（1）为了使用到以上功能，需要引入 Spring Validation 依赖。

```
<dependency>
    <groupId>org.springframework.boot</groupId>
    <artifactId>spring-boot-starter-validation</artifactId>
</dependency>
```

（2）使用注解标注需要验证的元素。以 SubmitArticleQuery.java 为例：

```
@Accessors(chain = true)
@Setter
@Getter
public class SubmitArticleQuery {
    //标题
    @NotNull(message = "Title must not be null.")
    private String title;
    //副标题
    @NotNull(message = "Headline must not be null.")
    private String headline;
    //正文
    @NotNull(message = "Content must not be null.")
    private String content;
    //作者名
    @NotNull(message = "Author must not be null.")
    private String author;
}
```

（3）使用@Validated 标注需要验证的入参。以 HtmlController.java 为例：

```
@PostMapping(value = "/article", headers = "Accept=application/xml",
produces = MediaType.APPLICATION_XML_VALUE)
```

```
    @ResponseBody
    public Article submitArticleAndGetXml(@RequestBody @Validated
SubmitArticleQuery query) {
        return submitArticle(query);
    }

    @PostMapping("/article")
    @ResponseBody
    public Article submitArticleAndGetJson(@RequestBody @Validated
SubmitArticleQuery query) {
        return submitArticle(query);
    }

    private Article submitArticle(SubmitArticleQuery query) {
        //用于接收 POST 请求
        User author = userRepository.findByLogin(query.getAuthor());
        if (author == null) {
            //返回 400 错误码
            throw new ResponseStatusException(HttpStatus.BAD_REQUEST, "This user
does not exist");
        }
        return articleService.saveArticle(query, author);
    }
```

该注解适用于程序的任意一层。在路径 src/main/java/com/example/myblog/service 下创建 ArticleService.java：

```
@Service
@RequiredArgsConstructor
public class ArticleService {

    private final ArticleRepository articleRepository;

    public Article saveArticle(@Validated SubmitArticleQuery query, User
author) {
        Article toSave = new Article().setAuthor(author)
                .setTitle(query.getTitle())
                .setHeadline(query.getHeadline())
                .setContent(query.getContent())
                .setSlug(CommonUtil.toSlug(query.getTitle()));
        articleRepository.save(toSave);
        return toSave;
    }

}
```

如此一来，一个基础的 Spring Validation 就实现了。可以观察到，使用不符合验证规则的内容进行请求，接口将会返回 400 错误以及验证相关的信息。在路径 src/main/te-st/com/example/myblog 下创建 ValidationTests.java：

```
@RunWith(SpringRunner.class)
@SpringBootTest
public class ValidationTests {

    @Autowired
```

```java
private WebApplicationContext wac;

@Autowired
private ObjectMapper mapper;

private MockMvc mockMvc;

@Before
public void setup() {
    mockMvc = MockMvcBuilders.webAppContextSetup(wac).build();
}

@Test
public void whenSubmitWrongArgument_thenReturn4xx() throws Exception {
    SubmitArticleQuery submitArticleQuery = new SubmitArticleQuery()
            .setTitle("title")
            .setHeadline("headline")
            //设置 content 为 Null 用于测试接口验证结果
            .setContent(null)
            .setAuthor("meimeihan");
    mockMvc.perform(
            MockMvcRequestBuilders
                    .post("/article")
                    .contentType(MediaType.APPLICATION_JSON)
                    .content(mapper.writeValueAsBytes(submitArticleQuery))
    ).andExpect(MockMvcResultMatchers.status().is4xxClientError())
            .andDo(print());
}
}
```

3.3.3 自定义校验

Spring Validation 提供了自定义校验的途径。通过实现 Validator 接口并定义对应的注解即可完成自定义校验。下面以验证 author 字段为例，实现一组自定义校验。

【示例3-8】

（1）创建自定义注解。在路径 src/main/java/com/example/myblog/validator 下创建 Author.java：

```java
@Target({FIELD})
@Retention(RUNTIME)
@Constraint(validatedBy = AuthorValidator.class)
@Documented
public @interface Author {

    String message() default "Author is not allowed.";

    Class<?>[] groups() default {};

    Class<? extends Payload>[] payload() default {};

}
```

（2）创建 Author 所需的 Validator。在路径 src/main/java/com/example/myblog/validator 下创建 AuthorValidator.java：

```java
public class AuthorValidator implements ConstraintValidator<Author, String> {

    private final List<String> VALID_AUTHORS = Arrays.asList("meimeihan", "leili");

    @Override
    public boolean isValid(String value, ConstraintValidatorContext context) {
        //判断验证是否通过的业务逻辑
        return VALID_AUTHORS.contains(value);
    }
}
```

（3）使用 Author 注解。

```java
@Accessors(chain = true)
@Setter
@Getter
public class SubmitArticleQuery {
    //标题
    @NotNull(message = "Title must not be null.")
    private String title;
    //副标题
    @NotNull(message = "Headline must not be null.")
    private String headline;
    //正文
    @NotNull(message = "Content must not be null.")
    private String content;
    //作者名
    @Author
    @NotNull(message = "Author must not be null.")
    private String author;
}
```

（4）测试自定义注解。

```java
@RunWith(SpringRunner.class)
@SpringBootTest
public class ValidationTests {
    //省略若干字段与方法
    @Test
    public void whenSubmitWrongAuthor_thenReturn4xx() throws Exception {
        SubmitArticleQuery submitArticleQuery = new SubmitArticleQuery()
                .setTitle("title")
                .setHeadline("headline")
                //设置 content 为 Null 用于测试接口验证结果
                .setContent("content")
                .setAuthor("anonymous");
        mockMvc.perform(
                MockMvcRequestBuilders
                        .post("/article")
                        .contentType(MediaType.APPLICATION_JSON)
```

```
                .content(mapper.writeValueAsBytes(submitArticleQuery))
        ).andExpect(MockMvcResultMatchers.status().is4xxClientError())
            .andDo(print());
    }
}
```

3.4 错误处理

人非圣贤，孰能无过。再精明强干的程序员编写的程序也会出现错误。在 Java 中，程序出现错误会抛出"不正常信息"（Throwable）。Throwable 又被分为"错误"（Error）和"异常"（Exception）。有别于人为失误造成的"故障"（Bug），异常在程序中代表的是出现了当前代码无法处理的状况。例如：在一个对象不存在（值为 Null）的情况下，调用该对象的某个方法引发了空指针；用户输入了一段 URL，但并没有找到对应的资源；一段计算过程中，0 被当作除数……完善的错误处理，使程序不会意外崩溃甚至能友好地提示用户进行正确操作，这是让程序变得愈发健壮的重要处理步骤。图 3.3 所示为 Spring Boot 的 Whitelabel Error Page。

Whitelabel Error Page

This application has no explicit mapping for /error, so you are seeing this as a fallback.

Thu Jun 27 18:33:27 CST 2019
There was an unexpected error (type=Not Found, status=404).
No message available

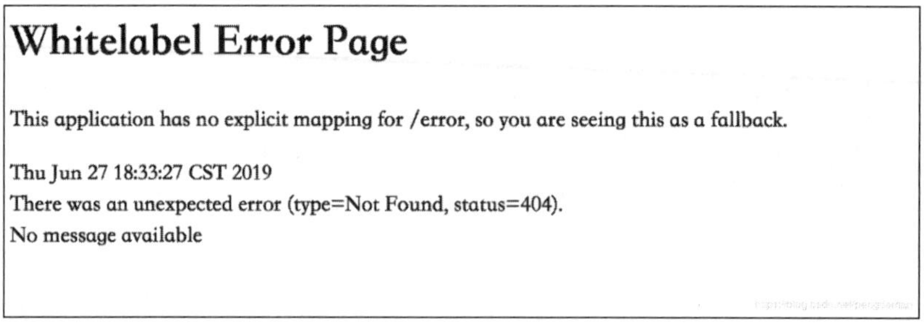

图 3.3　Whitelabel Error Page

在 Java 开发中，异常特别是检查型异常（Checked Exception），通常需要进行 try/catch 处理。而在基于 Spring Boot 的开发过程中，异常处理有了更多处理方式。

3.4.1　使用@ExceptionHandler 处理异常

首先要介绍的解决方案是使用注解@ExceptionHandler。该注解主要用于在 Controller 层面进行相同类型的异常处理。在对应 Controller 类中定义异常处理方法，并为其使用@ExcptionHandler 注解。Spring 将检测到该注解，把该方法注册为对应异常类及其子类的异常处理程序。异常处理的示例代码如下：

```
@ExceptionHandler({MyException.class})
public void handleException(MyException e) {
    //这里可以任意编写异常处理逻辑
    log.info("got an exception" + e.toString());
}
```

使用该注解的方法可以拥有非常灵活的签名，包括以下类型：

- 异常类型（Throwable）：可以选择一个大概的异常类型。例如，示例里的签名可以改为"Throwable e"或者"Exception e"，或者一个具体的异常类型。
- 请求与响应对象（Request/Response）：比如 ServletRequest/HttpServletRequest。
- InputStream/Reader：用于访问请求的内容。
- OutputStream/Writer：用户访问响应的内容。
- Model：作为从该方法返回 Model 的替代方案。

在返回类型方面也有灵活的选择：

- ModelAndView 对象。
- Model 对象，其对应视图由 RequestToViewNameTranslator 隐式确定。
- Map 对象，其对应视图同样由 RequestToViewNameTranslator 隐式确定。
- 值为某视图名的 String 对象，用于指定视图。
- 另外，在使用@ResponseBody 注解标识的情况下，将返回值使用转换器转换为最终结果。
- 使用 HttpEntity<?>或 ResponseEntity<?>对象包装的结果。
- 使用 void 作为返回类型，然后用签名中的 Response、OutputStream 或者 Writer 编写响应内容。

3.4.2　使用 HandlerExceptionResolver 处理异常

@ExceptionHandler 功能足够强大，但在不进行特殊处理的前提之下只能处理单个 Controller 的异常。面对多个 Controller 抛出的异常，还需要依赖 HandlerExceptionResolver 这一手段进行处理。使用 HandlerExceptionResolver 可以解决应用程序内的任何异常，并且依赖它可以实现 RESTful 服务的统一异常处理机制。

HandlerExceptionResolver 是一个公共接口。常见的使用方式是实现一个自定义的处理类。在自定义处理类之前，可以了解一下现有的部分实现：

- ExceptionHandlerExceptionResolver：该处理类就是帮助@ExceptionHandler 生效的组件。
- DefaultHandlerExceptionResolver：用于将标注的 Spring 异常解析为对应的 HTTP 状态码。
- ResponseStatusExceptionResolver：与注解@ResponseStatus 一起使用，将自定义异常与 HTTP 状态码进行对应，示例代码如下：

```
@ResponseStatus(value = HttpStatus.NOT_FOUND)
public class MyException extends Exception{

    public MyException() {
    }

    public MyException(String message) {
        super(message);
    }
}
```

之所以需要自定义处理类，原因在于以上实现无法控制响应体的内容。而大多数情况下，REST 服务的响应都需要有 JSON 或者 XML 格式的响应内容。

自定义处理类的示例代码如下:

```java
@Component
@Slf4j
public class RestResponseStatusExceptionResolver extends AbstractHandlerExceptionResolver {

    @Override
    protected ModelAndView doResolveException(
            HttpServletRequest request,
            HttpServletResponse response,
            Object handler,
            Exception ex) {
        try {
            if (ex instanceof IllegalArgumentException) {
                return handleIllegalArgument((IllegalArgumentException) ex, response, request);
            }
            //异常处理逻辑
        } catch (Exception handlerException) {
            log.warn("Handling of [" + ex.getClass().getName() + "]resulted in Exception", handlerException);
        }
        return null;
    }

    private ModelAndView handleIllegalArgument(IllegalArgumentException ex, HttpServletResponse response, HttpServletRequest request)
            throws IOException {
        response.sendError(HttpServletResponse.SC_CONFLICT);
        String accept = request.getHeader(HttpHeaders.ACCEPT);
        //处理响应内容
        return new ModelAndView();
    }
}
```

3.4.3 使用@ControllerAdvice 处理异常

在 Spring 3.2 版本引入了@ControllerAdvice 这一注解,为全局的@ExceptionHandler 提供了支持。将这个注解批注在一个处理类上,即可让该类下由@ExceptionHandler 批注的方法在全局层面对异常进行处理。

还记得上一小节参数验证失败的结果吗?请求中包含不符合要求的内容将抛出异常 MethodArgumentNotValidException。默认的响应内容如下:

```
{
    "timestamp": "2020-08-27T14:24:38.927+00:00",
    "status": 400,
    "error": "Bad Request",
    "trace": "org.springframework.web.bind.MethodArgumentNotValidException: ……",
    "message": "Validation failed for object='submitArticleQuery'. Error count:
```

```
1",
    "errors": [
        {
            "codes": [
                "NotNull.submitArticleQuery.content",
                "NotNull.content",
                "NotNull.java.lang.String",
                "NotNull"
            ],
            "arguments": [
                {
                    "codes": [
                        "submitArticleQuery.content",
                        "content"
                    ],
                    "arguments": null,
                    "defaultMessage": "content",
                    "code": "content"
                }
            ],
            "defaultMessage": "Content must not be null.",
            "objectName": "submitArticleQuery",
            "field": "content",
            "rejectedValue": null,
            "bindingFailure": false,
            "code": "NotNull"
        }
    ],
    "path": "/article"
}
```

其中 trace 属性将会输出大段落的堆栈信息，在此处做了省略处理。对于调用方而言，返回的信息或许存在冗余。可以根据需求使用该方案对其进行调整。

在路径 src/main/java/com/example/myblog/controller 下创建 MyBlogControllerAdvice.java：

```
@ControllerAdvice
@Slf4j
public class MyBlogControllerAdvice {
    @ResponseBody
    @ExceptionHandler(value = MethodArgumentNotValidException.class)
    public Result<String> errorHandler(MethodArgumentNotValidException e) {
        String errorMsg =
e.getBindingResult().getAllErrors().get(0).getDefaultMessage();
        log.error("未处理异常" + errorMsg);
        return new Result<String>()
                .setMessage("参数错误: " + errorMsg);
    }
}
```

响应结果如下：

```
{
    "code": 0,
    "data": null,
    "message": "参数错误: Content must not be null."
```

}
```

### 3.4.4　抛出 ResponseStatusException 异常

上文介绍的方法大多适用于解决一个切面上的问题。如果仅需要针对少数接口进行异常处理，控制其返回给客户端的 HTTP 状态码、错误指引以及报错原因，那么 ResponseStatusException 将会是一个不错的选择。示例代码如下：

```java
@GetMapping(value = "/{id}")
public Foo findById(@PathVariable("id") Long id, HttpServletResponse response)
{
    try {
        Foo resourceById = RestPreconditions.checkFound(service.findOne(id));
        eventPublisher.publishEvent(new SingleResourceRetrievedEvent(this, response));
        return resourceById;
    }
    catch (MyResourceNotFoundException exc) {
        throw new ResponseStatusException(
          HttpStatus.NOT_FOUND, "Foo Not Found", exc);
    }
}
```

3.5　Swagger 文档

在前后端分离的软件架构模式下，前后端程序的开发人员大多不是同一个人。这时，接口文档的重要性便体现出来了。接口文档在项目初期帮助前端开发人员快速理解接口功能，在项目后期方便维护人员对服务进行查看与维护。一份内容详实的接口文档，可以为开发带来巨大的便利，相应地也要耗费不少心血，毕竟单是为了保持接口与文档的版本一致，所付出的努力都是难以忽视的。

相信不少前后端开发工程师都或多或少被接口文档"折磨"过。前端开发抱怨文档不够友好，后端开发烦恼于文档工作过于耗时、耗力。不过在建立了规范并将流程自动化之后，接口文档将不再是难题。本节将介绍如何在 Spring Boot 中集成一款自动生产 API 文档的工具——Swagger。

3.5.1　Swagger/OpenAPI 规范

Swagger 是一个开源项目，主要用于 RESTful API 的描述与调试。集成了一组 HTML、JavaScript 和 CSS 前端资源，从符合 Swagger 规范的 API 中动态生成可以交互的接口文档。在过去几年中，Swagger2 已经成为了定义或记录 API 的一种规范。之后，该规范被移至 Linux 基金会，并重新命名为 OpenAPI 规范。下文对 Swagger/OpenAPI 的描述，事实上指代的是同一事物。

为了在项目中整合 Swagger，首先需要添加 Swagger 的 starter 依赖。该依赖由 Spring 社区内一个非官方组织 Springfox 所维护，使用该 starter 可以方便地整合 Swagger。依赖配置如下：

```xml
<dependency>
```
```

```
        <groupId>io.springfox</groupId>
        <artifactId>springfox-boot-starter</artifactId>
        <version>3.0.0</version>
</dependency>
```

3.5.2 生成接口文档

接口文档的生成，依赖于 Docket。为了生成一份接口文档，需要在 Spring Boot 程序中配置一个 Docket 的 Java Bean。得益于 Swagger 的合理设计，需要配置的内容并不复杂。在路径 /src/main/java/com/example/myblog/config 下创建 SwaggerConfig.java：

【示例 3-9】

```
@EnableOpenApi
@Configuration
public class SwaggerConfig {
    @Bean
    public Docket createRestApi() {
        return new Docket(DocumentationType.OAS_30)
                .apiInfo(apiInfo())
                .select()
                //使用@ApiOperaion 的 Controller 将被添加至接口文档当中
                .apis(RequestHandlerSelectors.withMethodAnnotation(ApiOperation.class))
                .paths(PathSelectors.any())
                .build();
    }

    private ApiInfo apiInfo() {
        return new ApiInfoBuilder()
                .title("Swagger3 接口文档")
                .description("Swagger 整合示例")
                .version("1.0")
                .build();
    }
}
```

创建好 Swagger 配置之后，可以分别在路径 http://localhost:8080/v2/api-docs 与 http://localhost:8080/v3/api-docs 下访问到 Swagger2 规范的接口文档与 OpenAPI3 的接口文档。

http://localhost:8080/v2/api-docs 初始内容：

```
{
  "swagger": "2.0",
  "info": {
    "description": "Swagger 整合示例",
    "version": "1.0",
    "title": "Swagger3 接口文档"
  },
  "host": "localhost:8080",
  "basePath": "/"
}
```

http://localhost:8080/v3/api-docs 初始内容：

```
{
  "openapi": "3.0.3",
  "info": {
    "title": "Swagger3 接口文档",
    "description": "Swagger 整合示例",
    "version": "1.0"
  },
  "servers": [
    {
      "url": "http://localhost:8080",
      "description": "Inferred Url"
    }
  ],
  "components": {
  }
}
```

另外，在 http://localhost:8080/swagger-ui/index.html 下可以访问到用来与后端交互的接口文档界面。不过文档暂时还是空空如也，等待开发人员通过后面的操作将文档渐渐丰富起来。Swagger3 的接口文档界面如图 3.4 所示。

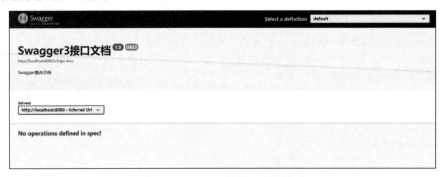

图 3.4　Swagger3 的接口文档界面

3.5.3　使用注解生成文档内容

无论是 api-doc 还是 swagger-ui，这两者的内容都依赖于开发者在代码中通过 Swagger 提供的注解进行完善。

（1）@Api：用于 Controller 类上，将该类标记为 Swagger 的资源。常用参数如下：

- tags：说明该类的作用，参数类型为 String 数组。

（2）@ApiOperation：用于接口方法上，描述针对特定路径下的操作。常用参数如下：

- value：方法的用途和作用。
- notes：方法的注意事项和备注。

（3）@ApiModel：用于实体类上，描述实体作用。常用参数如下：

- description：描述实体的作用。

（4）@ApiModelProperty：用于实体属性上，描述实体的属性。常用参数如下：
- value：对属性的简要描述。
- name：属性名。
- required：参数是否是必选的。

（5）@ApiImplicitParam：用于方法，描述隐含的参数。常用参数如下：
- name：参数名。
- value：参数说明。
- dataType：数据类型。
- paramType：用于描述参数的类型（参数所处位置）。可选内容包括：path、query、body、header、form。

（6）@ApiImplicitParams：用于方法上，包含多个@ApiImplicitParam。

（7）@ApiParam：用于方法、参数，用于描述请求的要求和说明。常用参数如下：
- name：参数名。
- value：对参数的简要描述。
- defaultValue：参数默认值。
- required：参数是否是必选的。

（8）@ApiResponse：用于请求的方法上，描述不同的响应。常用参数如下：
- code：表示响应的状态码。
- message：描述状态码对应的响应信息。

（9）@ApiResponses：用于方法上，包含多个@ApiResponse。

示例代码 SimpleRestController.java：

```java
@Api(tags = "RESTful 服务传参示例")
@RestController
@RequestMapping("/api/simple")
public class SimpleRestController {

    @ApiOperation(value = "无参 GET 请求", notes = "用于演示通过无参 GET 请求的形式对接口进行请求")
    @GetMapping("/no-param")
    public Result<String> noParam() {
        //无参
        return Result.ok("No parameter.");
    }

    @ApiOperation(value = "单个参数的 GET 请求", notes = "用于演示通过单参数 GET 请求的形式对接口进行请求")
    @ApiImplicitParam(name = "implicit", value = "提供隐含参数的输入方式")
    @GetMapping(path = "/single-param", params = "implicit")
    public Result<String> singleParam(@ApiParam(name = "单参数") String param) {
        //单个可选参数
        return Result.ok("The parameter is :" + param);
    }
```

```java
    @ApiOperation(value = "下一个生日", notes = "输入出生年、月、日，计算到下一个生
日的天数")
    @ApiResponses({
            @ApiResponse(code = 400, message = "输入日期大于当前日期"),
            @ApiResponse(code = 200, message = "成功")
    })
    @GetMapping("/next-birth-day/{year}/{month}/{day}")
    public ResponseEntity<Result<Long>> nextBirthday(@PathVariable int year,
@PathVariable int month, @PathVariable int day) {
        LocalDate birthDate = LocalDate.of(year, month, day);
        if (birthDate.isAfter(LocalDate.now())) {
            return new ResponseEntity<>(null, HttpStatus.BAD_REQUEST);
        }
        LocalDate nextBirthDay = LocalDate.of(LocalDate.now().getYear(),
birthDate.getMonth(), birthDate.getDayOfMonth());
        if (nextBirthDay.isBefore(LocalDate.now())) {
            nextBirthDay = nextBirthDay.plusYears(1);
        }
        return new ResponseEntity<>(Result.ok(DAYS.between(LocalDate.now(),
nextBirthDay)), HttpStatus.OK);
    }
    //省略若干方法……
}
```

示例代码 Result.java：

```java
@ApiModel(description = "用于统一 RESTful 服务返回内容")
@Accessors(chain = true)
@Setter
@Getter
public class Result<T> {
    @ApiModelProperty("业务状态码")
    private int code;
    @ApiModelProperty("响应内容")
    private T data;
    @ApiModelProperty("错误信息")
    private String message;

    private final static int SUCCESS = 0;
    private final static int FAIL = -1;

    public static <T> Result<T> ok(T data) {
        return new Result<T>()
                .setCode(SUCCESS)
                .setData(data);
    }

    public static <T> Result<T> failed(String message) {
        return new Result<T>()
                .setCode(FAIL)
                .setMessage(message);
    }
}
```

第 4 章

数据库与持久化技术

持久化,意味着数据从瞬时状态转化为持久状态。没有持久化这一步骤,数据将仅存在于内存(RAM)当中。受限于内存的特性,内存断电时其中的数据将会一并丢失。当遇到需要存储于硬盘的数据时,将要借助数据持久化技术来实现。Spring Boot 应用程序开发过程中所用到的数据持久化,大多依赖于各种类型的数据库、对应的客户端以及相关持久化框架。

本章将围绕数据库与持久化技术这一主题展开,主要涉及的知识点有:

- 常见关系型数据库与非关系型数据库的特性以及使用场景
- 使用 JDBC 与 JPA 操作关系型数据库
- 使用 JPA 操作非关系型数据库
- Redis 的使用

4.1 使用 JdbcTemplate 访问关系型数据库

JDBC(Java Database Connectivity)是 Java 提供用于编写应用程序作为客户端访问数据库的 API,JdbcTemplate 则是 Spring 对 JDBC 的封装。它简化了 JDBC 的使用并有助于避免常见错误。JdbcTemplate 是一种轻量的持久化解决方案,仅需要非常简单的配置就可以实现对数据库的访问。因为是轻量的解决方案,当请求数据库的需求并不复杂时,JdbcTemplate 将会是一个非常好的选择。

4.1.1 引入依赖

为了启用 JdbcTemplate,需要引入依赖 spring-boot-starter-jdbc。另外本章以 MySQL 为例,因此还需要引入 mysql-connector-java 的依赖:

```xml
<dependency>
    <groupId>org.springframework.boot</groupId>
    <artifactId>spring-boot-starter-jdbc</artifactId>
</dependency>
<dependency>
    <groupId>mysql</groupId>
    <artifactId>mysql-connector-java</artifactId>
</dependency>
```

4.1.2 准备数据

Spring Boot 提供了一种自动初始化数据库的方式：从 classpath 中加载 schema.sql 与 data.sql，两个 sql 脚本分别用于创建表结构和初始化表数据。如果初始化数据需要分别支持不同的数据库，可以使用配置项 spring.datasource.platform 区分当前环境所连接的平台，并且通过"schema-${platform}.sql"与"data-${platform}.sql"这种命名方式创建对应平台的脚本文件。

这一初始化行为通过配置项 spring.datasource.initialization-mode 来控制，可选的参数有：

```
//总是执行初始化操作
DataSourceInitializationMode.ALWAYS
//仅当数据源为嵌入式数据库的时候
DataSourceInitializationMode.EMBEDDED
//从不执行初始化操作
DataSourceInitializationMode.NEVER
```

在默认情况下，Spring Boot 启用 Spring JDBC 初始化程序的快速失败功能。这意味着如果这两个脚本的执行出现异常，就会导致程序无法启动。可以通过配置项 spring.datasource.continue-on-error 做出调整。

application.yml 示例如下：

```yaml
#application.yml 模板
spring:
  datasource:
    url: jdbc:mysql://127.0.0.1:3306/dev?useSSL=false
    username: root
    password: dbpass
    #当执行 schema.sql 的用户不同时，可以配置以下两项
    schema-username: root
    schema-password: dbpass
    #当执行 data.sql 的用户不同时，可以配置以下两项
    data-username: root
    data-password: dbpass
    platform: mysql
    schema: classpath:schema.sql
    data: classpath:data.sql
    continue-on-error: true
    initialization-mode: always
```

schema.sql：

```
-- ----------------------------
-- Table structure for vehicle
-- ----------------------------
```

```sql
DROP TABLE IF EXISTS `vehicle`;

CREATE TABLE `vehicle` (
  `id` int(11) NOT NULL AUTO_INCREMENT,
  `name` varchar(255) DEFAULT NULL,
  `price` decimal(10,2) DEFAULT NULL,
  PRIMARY KEY (`id`)
)
ENGINE=InnoDB DEFAULT CHARSET=utf8mb4;
```

data.sql：

```sql
-- ----------------------------
-- Records of vehicle
-- ----------------------------
INSERT INTO `vehicle` VALUES (1, 'Bentley', 2750000.00);
INSERT INTO `vehicle` VALUES (2, 'Land Rover', 1468000.00);
INSERT INTO `vehicle` VALUES (3, 'Porsche', 860000.00);
INSERT INTO `vehicle` VALUES (4, 'Mercedes', 1980000.00);
INSERT INTO `vehicle` VALUES (5, 'BMW', 561800.00);
INSERT INTO `vehicle` VALUES (6, 'InfinitI', 798000.00);
INSERT INTO `vehicle` VALUES (7, 'Cadillac', 1380000.00);
INSERT INTO `vehicle` VALUES (8, 'Lincoln', 1580000.00);
```

4.1.3 queryForObject()方法

首先介绍基础的 queryForObject()方法。该方法会执行一条 SQL 语句并得到一个结果对象，其中结果对象的类型需要在参数中声明。示例代码如下：

```java
@Transactional(rollbackFor = Exception.class)
@SpringBootTest
class JdbcApplicationTests {

    @Autowired
    private JdbcTemplate jdbcTemplate;

    @Test
    void queryForObject() {
        //查询 vehicle 表记录数
        String sql = "SELECT count(*) FROM vehicle";
        //获得 SQL 语句的执行结果
        Integer numOfVehicle = jdbcTemplate.queryForObject(sql, Integer.class);
        assert numOfVehicle != null;
        System.out.format("There are %d vehicles in the table", numOfVehicle);
    }
}
```

4.1.4 使用 RowMapper 映射实体

当以上示例中的返回类型由包装类修改为自定义的类型时，或许会碰到 IncorrectResult

SetColumnCountException 这一异常。这是因为该方法并不支持自动化映射操作,如果需要将结果与自定义类型进行映射,将需要 RowMapper 的帮助。通过编写一个 RowMapper 的具体实现以完成结果到自定义类型的映射。

首先创建与数据库结构所对应的实体类 Vehicle.java:

```
@NoArgsConstructor
@Accessors(chain = true)
@Setter
@Getter
@ToString
public class Vehicle {
    private Integer id;
    private String name;
    private BigDecimal price;
}
```

RowMapper 相关示例代码如下:

```
@Transactional(rollbackFor = Exception.class)
@SpringBootTest
class JdbcApplicationTests {

    @Autowired
    private JdbcTemplate jdbcTemplate;

    @Test
    void queryForObject_WithRowMapper() {
        RowMapper<Vehicle> rm = (ResultSet result, int rowNum) -> new Vehicle()
                .setId(result.getInt("id"))
                .setName(result.getString("name"))
                .setPrice(result.getBigDecimal("price"));
        //使用"?"作为占位符,执行 SQL 时将会将其替换成对应的参数
        String sql = "SELECT id,name,price FROM vehicle WHERE id = ?";
        int id = 1;
        Vehicle vehicle = jdbcTemplate.queryForObject(sql, new Object[]{id}, rm);
        assert vehicle != null;
        System.out.println(vehicle.toString());    }
}
```

4.1.5 使用 BeanPropertyRowMapper 映射

RowMapper 固然很不错,但如果每查询出一组新结果需要进行映射操作,就需要实现一个 RowMapper,那么这个开发体验就不够友好。好在 BeanPropertyRowMapper 打消了这种顾虑。当查询结果的字段名与映射类的属性名一致时,可以使用 BeanPropertyRowMapper 代替手工实现 RowMapper。

> **注 意**
>
> 映射类需要有默认或者无参构造器。

示例代码如下：

```
@Transactional(rollbackFor = Exception.class)
@SpringBootTest
class JdbcApplicationTests {

    @Autowired
    private JdbcTemplate jdbcTemplate;

    @Test
    void queryForObject_WithBeanPropertyRowMapper() {
        String sql = "SELECT id,name,price FROM vehicle WHERE id = ?";
        int id = 1;
        //使用BeanPropertyRowMapper代替手动实现RowMapper接口
        Vehicle vehicle = jdbcTemplate.queryForObject(sql, new Object[]{id},
BeanPropertyRowMapper.newInstance(Vehicle.class));
        assert vehicle != null;
        System.out.println(vehicle.toString());
    }
}
```

4.1.6　queryForList()方法

以上示例的查询结果都是单个对象，当查询结果类型为列表时，则需要用到 queryForList()方法。示例代码如下：

```
@Transactional(rollbackFor = Exception.class)
@SpringBootTest
class JdbcApplicationTests {

    @Autowired
    private JdbcTemplate jdbcTemplate;

    @Test
    void queryForList() {
        String sql = "SELECT id,name,price FROM vehicle";
        List<Map<String, Object>> ret = jdbcTemplate.queryForList(sql);
        assert !ret.isEmpty();
        ret.forEach(System.out::println);
    }
}
```

4.1.7　不同的 JDBCTemplate 实现 NamedParameterJdbcTemplate

使用基础的 JdbcTemplate 执行 SQL 语句时，所编写的 SQL 语句的参数占位符默认为"?"。当需要传入的参数增加到一定数量，参照这样编写的 SQL 语句将会变得难以理解。此时将需要用到 NamedParameterJdbcTemplate。使用 NamedParameterJdbcTemplate 可以将有含义的占位符来代替"?"。示例代码如下：

```
@Transactional(rollbackFor = Exception.class)
@SpringBootTest
```

```java
class JdbcApplicationTests {

    @Autowired
    private NamedParameterJdbcTemplate namedParameterJdbcTemplate;

    @Test
    void queryForObject_WithNamedParameterJdbcTemplate() {
        String sql = "SELECT id,name,price FROM vehicle WHERE name like :name AND price > :price LIMIT 1";
        //参数名与 SQL 语句中的占位符需要保持一致
        MapSqlParameterSource mapSqlParameterSource = new MapSqlParameterSource()
                .addValue("name", "B%")
                .addValue("price", "60000");
        Vehicle vehicle = namedParameterJdbcTemplate.queryForObject(sql,
                mapSqlParameterSource, BeanPropertyRowMapper.newInstance(Vehicle.class));
        assert vehicle != null;
        System.out.println(vehicle.toString());
    }
}
```

4.1.8 update()方法

以上示例都在介绍如何获取一个或一组数据，当需要对记录进行更新时，则需要调用 update() 方法。该更新操作包含对记录的增加、修改以及删除。示例代码如下：

```java
@Transactional(rollbackFor = Exception.class)
@SpringBootTest
class JdbcApplicationTests {

    @Autowired
    private NamedParameterJdbcTemplate namedParameterJdbcTemplate;

    @Test
    void update_SaveVehicle() {
        //新增一条记录
        String sql = "INSERT INTO vehicle(name,price) VALUES(:name,:price)";
        MapSqlParameterSource mapSqlParameterSource = new MapSqlParameterSource()
                .addValue("name", "Tesla")
                .addValue("price", "850000");
        int ret = namedParameterJdbcTemplate.update(sql,
                mapSqlParameterSource);
        assert ret > 0;
    }

    @Test
    void update_UpdateVehicle() {
        //更新数条记录
        String sql = "UPDATE vehicle SET price = price * 0.9 WHERE price > :price";
        MapSqlParameterSource mapSqlParameterSource = new MapSqlParameterSource()
                .addValue("price", "1000000");
```

```
            int ret = namedParameterJdbcTemplate.update(sql,
mapSqlParameterSource);
        assert ret > 0;
    }

    @Test
    void update_DeleteVehicle() {
        //删除数条记录
        String sql = "DELETE FROM vehicle WHERE price < :price";
        MapSqlParameterSource mapSqlParameterSource = new
MapSqlParameterSource()
                .addValue("price", "1000000");
        int ret = namedParameterJdbcTemplate.update(sql,
mapSqlParameterSource);
        assert ret > 0;
    }
}
```

4.2 JPA 与关系型数据库

上一节介绍了使用 JDBCTemplate 访问关系型数据库的方式。其特点在于简单直接，适合规模不大的 Web 项目。本节将介绍更适合大型项目访问关系型数据库的方式——JPA，使用它可以轻松构建出更为复杂的系统。选择哪种访问方式，需要根据项目的场景来决定合适的解决方案，任何技术都不可以被当作"银弹"来使用。

4.2.1 什么是 JPA

JPA（Java Persistence API）是在 JDK 1.5 之后提出的 Java 持久化规范（JSR 338）。JPA 的出现是为了简化 Web 程序的持久层开发并且对当时 ORM 技术进行整合，结束当时所流行的各种 ORM 框架，如 Hibernate、TopLink 以及 JDO，各自为营的局面。

JPA 的操作步骤：

- 加载配置文件，并根据配置创建实体管理工厂对象。
- 使用实体管理工厂对象创建实体管理器。
- 创建事务对象并开启事务。
- CURD 操作。
- 提交事务。
- 释放资源。

示例代码如下：

```
// 1.加载配置文件，并根据配置创建实体管理工厂对象
    EntityManagerFactory factory =
Persistence.createEntityManagerFactory("jpaConfig");
    // 2.使用实体管理工厂对象创建实体管理器
```

```
EntityManager em = factory.createEntityManager();
// 3.获取事务对象，开启事务
EntityTransaction tx = em.getTransaction();
tx.begin();// 开启事务
// 4.完成增删改查操作
User user = new User();
user.setName("张三");
// 保存
em.persist(customer);
// 5.提交事务
tx.commit();
// 6.释放资源
em.close();
factory.close();
```

整体架构如图 4.1 所示。

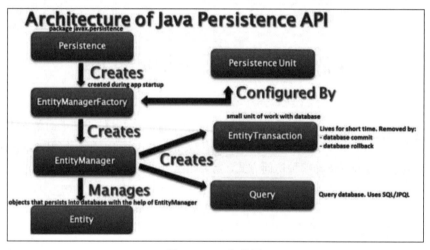

图 4.1　JPA 架构图

4.2.2　再谈 Spring Data JPA

Spring Data JPA 是 Spring Data 大家族中的一员，可以轻松实现基于 JPA 的存储库。Spring Data JPA 主要基于 JPA 提供对数据访问层的增强支持。借助它可以使构建设计数据访问技术的 Spring 应用程序变得更加容易。

在相当长的一段时间内，实现应用程序的数据访问层一直很麻烦。必须编写大量的样板代码来执行简单查询以及执行分页和审计。Spring Data JPA 旨在通过减少实际需要的工作量来显著改善数据访问层的实现。作为开发人员可以只编写 Repository 接口，包括自定义查找器方法，Spring 将自动提供对应的实现。

Spring Data 生态如图 4.2 所示。

第 4 章 数据库与持久化技术 | 105

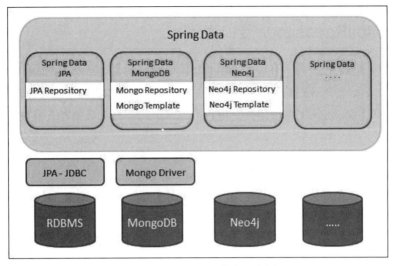

图 4.2 Spring Data 生态示意图

Spring Data JPA 是 Spring Data 对 JPA 规范的封装，在其规范下提供 Repository 层的实现，并提供配置项用以切换具体实现规范的 ORM 框架。Spring Data JPA、JPA 以及基于 JPA 规范的 ORM 框架，如图 4.3 所示。

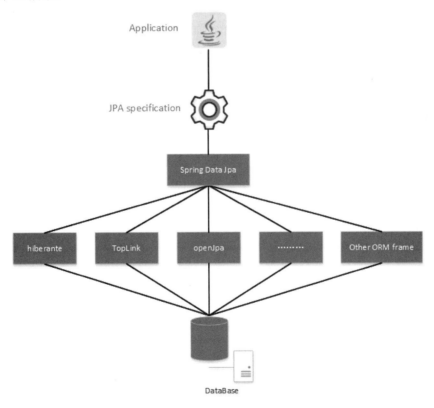

图 4.3 Spring Data JPA、JPA 与各 ORM 框架

4.2.3 基于 JpaRepository 接口查询

Spring Data JPA 框架的目标之一就在于简化数据访问层的开发过程，消除项目中的样板代码。基于接口的查询方式是实现代码简化的有效方法。通过基于接口的查询方式进行开发，框架将在应用运行时，根据接口名的定义生成包含对应 SQL 语句的代理实例。这免去了手写 SQL 的环节，进而实现了简化。

基于接口查询，首先要关注的接口为 "Repository"。Repository 是 Spring Data JPA 的核心接口。它需要领域实体类以及实体类的 ID 类型作为类型参数进行管理。该类主要作为标记接口，用以捕获要使用的类型并帮助发现扩展该接口的子接口。另外还有更为具体的 CrudRepository 以及 JpaRepository，这两个类包含具体的基础 CURD 方法。

【示例 4-1】

Repository.java：

```java
@Indexed
public interface Repository<T, ID> {

}
```

JpaRepository.java：

```java
@NoRepositoryBean
public interface JpaRepository<T, ID> extends PagingAndSortingRepository<T, ID>,
    QueryByExampleExecutor<T> {

    //查询所有数据
    @Override
    List<T> findAll();

    //查询所有数据，并以排序选项进行排序后返回
    @Override
    List<T> findAll(Sort sort);

    //根据 id 查询集合
    @Override
    List<T> findAllById(Iterable<ID> ids);

    //保存所有数据
    @Override
    <S extends T> List<S> saveAll(Iterable<S> entities);

    //将之前的改动刷写进数据库
    void flush();

    //保存并立刻刷写当前实体
    <S extends T> S saveAndFlush(S entity);

    //删除给出的集合
    void deleteInBatch(Iterable<T> entities);
```

```java
    //批量删除
    void deleteAllInBatch();

    //根据id查询目标实体
    T getOne(ID id);

    //根据实例查询
    @Override
    <S extends T> List<S> findAll(Example<S> example);

    //根据实例查询并排序
    @Override
    <S extends T> List<S> findAll(Example<S> example, Sort sort);
}
```

使用基于接口的查询方式，首先需要定义查询表对应的实体。实体示例 Patient.java 如下：

```java
@Entity
@Table(name = "patient")
@Data
@Accessors(chain = true)
public class Patient implements Serializable {

    private static final long serialVersionUID = 1L;

    /**
     * 主键
     */
    @Id
    @Column(name = "id", nullable = false)
    private Integer id;

    /**
     * 名
     */
    @Column(name = "first_name")
    private String firstName;

    /**
     * 姓
     */
    @Column(name = "last_name")
    private String lastName;

    /**
     * 身高
     */
    @Column(name = "height")
    private BigDecimal height;

    /**
     * 体重
     */
    @Column(name = "body_weight")
```

```
    private BigDecimal bodyWeight;

    /**
     * BMI 指数
     */
    @Column(name = "BMI")
    private BigDecimal DMI;

}
```

通过注解@Entity 标注该类为实体类，通过@Table(name = "patient")标注该类的表明为 patient。类的主键需要使用@Id 注解进行标注，另外需要@Column 注解标注对应的字段名。

实体查询类 PatientRepository.java 如下：

```
public interface PatientRepository extends JpaRepository<Patient, Integer>{

    List<Patient> findByFirstName(String firstName);

    Patient findByFirstNameAndLastName(String firstName, String lastName);

    List<Patient> findByHeightGreaterThan(BigDecimal height);
}
```

PatientRepository.java 对应测试代码如下：

```
@Test
public void testJpaRepository() {
    List<Patient> san = patientRepository.findByFirstName("san");
    assert san != null;
    assert san.size() > 0;
    Patient lisi = patientRepository.findByFirstNameAndLastName("si", "li");
    assert lisi != null;
    List<Patient> tallPatients =
patientRepository.findByHeightGreaterThan(new BigDecimal(190));

    assert tallPatients != null;
    assert tallPatients.size() == 0;
}
```

从以上示例可以观察到，完成一个查询仅需要定义一个 findBy{:column}格式的方法名。事实上，findBy 可以替换为 getBy、readBy 或者直接去掉。Spring Data JPA 将在应用运行时对方法名进行解析，解析的过程为：去掉 findBy 等前缀，再根据剩下的字段名与关键字，生成对应查询的代码实现。

关键字及示例参考表 4.1 所示。

表 4.1　关键字及示例参考

关键字	示例	JPQL 语句片段
And	findByLastnameAndFirstname	… where x.lastname = ?1 and x.firstname = ?2
Or	findByLastnameOrFirstname	… where x.lastname = ?1 or x.firstname = ?2
Is,Equals	findByFirstname,findByFirstnameIs,findByFirstnameEquals	… where x.firstname = ?1

（续表）

关键字	示例	JPQL 语句片段
Between	findByStartDateBetween	… where x.startDate between ?1 and ?2
LessThan	findByAgeLessThan	… where x.age < ?1
LessThanEqual	findByAgeLessThanEqual	… where x.age <= ?1
GreaterThan	findByAgeGreaterThan	… where x.age > ?1
GreaterThanEqual	findByAgeGreaterThanEqual	… where x.age >= ?1
After	findByStartDateAfter	… where x.startDate > ?1
Before	findByStartDateBefore	… where x.startDate < ?1
IsNull	findByAgeIsNull	… where x.age is null
IsNotNull,NotNull	findByAge(Is)NotNull	… where x.age not null
Like	findByFirstnameLike	… where x.firstname like ?1
NotLike	findByFirstnameNotLike	… where x.firstname not like ?1
StartingWith	findByFirstnameStartingWith	… where x.firstname like ?1(parameter bound with appended %)
EndingWith	findByFirstnameEndingWith	… where x.firstname like ?1(parameter bound with prepended %)
Containing	findByFirstnameContaining	… where x.firstname like ?1(parameter bound wrapped in %)
OrderBy	findByAgeOrderByLastnameDesc	… where x.age = ?1 order by x.lastname desc
Not	findByLastnameNot	… where x.lastname <> ?1
In	findByAgeIn(Collection\<Age> ages)	… where x.age in ?1
NotIn	findByAgeNotIn(Collection\<Age> age)	… where x.age not in ?1
TRUE	findByActiveTrue()	… where x.active = true
FALSE	findByActiveFalse()	… where x.active = false
IgnoreCase	findByFirstnameIgnoreCase	… where UPPER(x.firstame) = UPPER(?1)

4.2.4 基于 JpaSpecificationExecutor 接口查询

上一小节介绍的 JpaRepository 接口固然十分方便，但用于实现逻辑更为复杂的需求，便显得捉襟见肘了。使用 JpaRepository 接口更适用于参数不多、逻辑简单的查询场景。为了补足 JpaRepository 难以实现的部分，Spring Data JPA 另外提供了 JpaSpecificationExecutor 这一接口供复杂查询的场景使用。

【示例 4-2】

JpaSpecificationExecutor.java：

```
public interface JpaSpecificationExecutor<T> {
    //根据 spec 查询出一个 Optional 的实体类
    Optional<T> findOne(@Nullable Specification<T> spec);
```

```
    //根据 spec 查询出对应实体列表
    List<T> findAll(@Nullable Specification<T> spec);

    //根据 spec 查询出实体分页
    Page<T> findAll(@Nullable Specification<T> spec, Pageable pageable);

    //根据 spec 查询出对应实体列表，并根据给出的排序条件进行排序
    List<T> findAll(@Nullable Specification<T> spec, Sort sort);

    //查询满足 spec 条件的实体列表长度
    long count(@Nullable Specification<T> spec);
}
```

其中 Specification 接口提供的 toPredicate()方法，供开发人员灵活构造复杂的查询条件。Specification.java：

```
public interface Specification<T> extends Serializable {

    @Nullable
    Predicate toPredicate(Root<T> root, CriteriaQuery<?> query,
CriteriaBuilder criteriaBuilder);

    //省略若干方法
}
```

在基于接口查询的开发过程中，往往会在实体的 Repository 接口类同时继承 JpaRepository 与 JpaSpecificationExecutor，以赋予该 Repository 接口能同时完成简单查询与复杂查询的能力。

UserRepository.java：

```
public interface UserRepository extends JpaRepository<User, Integer>,
JpaSpecificationExecutor<User> {

}
```

实体类 User.java：

```
@Entity
@Table(name = "user")
@Data
@Accessors(chain = true)
public class User implements Serializable {

    private static final long serialVersionUID = 1L;

    /**
     * 主键
     */
    @Id
    @Column(name = "id", nullable = false)
    private Integer id;

    /**
     * 用户名
     */
    @Column(name = "name")
```

```java
    private String name;

    /**
     * 账户名
     */
    @Column(name = "account")
    private String account;

    /**
     * 密码
     */
    @Column(name = "password")
    private String password;

    /**
     * 创建时间
     */
    @Column(name = "create_time")
    private LocalDateTime createTime;

    @OneToOne
    private Vehicle vehicle;
}
```

设想一个场景，需要创建一个查询方法用于查询符合条件的用户。例如，给出一个时间区间与一个关键字，查询出创建时间在该区间内的所有用户，并且用户名包含关键字。如果使用 JpaRepository 实现，则示例代码如下：

```java
@Test
public void testJpaRepositoryComplicated() {
    //根据时间与关键字查询
    List<User> queryWithTimeAndKeyWord = getUser(LocalDateTime.of(2019, 1, 1, 0, 0, 0), LocalDateTime.of(2020, 1, 1, 0, 0, 0),
            "陈");
    assert queryWithTimeAndKeyWord != null;
    //根据时间查询
    List<User> queryWithTime = getUser(LocalDateTime.of(2019, 1, 1, 0, 0, 0),
            LocalDateTime.of(2020, 1, 1, 0, 0, 0),
            null);
    assert queryWithTime != null;
    //普通查询
    List<User> query = getUser(null, null, "张");
    assert query != null;
}

//以下示例为不推荐的查询实现方式，属于错误示范
private List<User> getUser(@Nullable LocalDateTime start, @Nullable LocalDateTime end, @Nullable String keyWord) {
    //String.format 中%为特殊字符需要再加一个%进行转义，以下格式化的结果为%{keyWord}%
    String nameLike = keyWord == null ? null : String.format("%%%s%%", keyWord);
    if (start != null && end != null && !StringUtils.isEmpty(nameLike)) {
        //查询条件同时包含时间与关键字
        return userRepository.findByCreateTimeBetweenAndNameLike(start, end, nameLike);
    } else if (start != null && end != null && StringUtils.isEmpty(nameLike))
```

```
{
        //查询条件仅包含时间
        return userRepository.findByCreateTimeBetween(start, end);
    } else if ((start == null || end == null) && !StringUtils.isEmpty(nameLike))
{
        //查询条件仅包含关键字
        return userRepository.findByNameLike(keyWord);
    } else {
        return userRepository.findAll();
    }
}
```

可以看到，这一段代码的实现并不优雅，需要针对不同情况定义不同的 Repository 接口。如果参数进一步增加，对应 Repository 接口内的方法数量将膨胀到难以维护的程度。使用 JpaSpecificationExecutor 实现同样的功能，示例代码如下：

```
@Test
public void testJpaSpecificationExecutor() {
    //根据时间与关键字查询
    List<User> queryWithTimeAndKeyWord =
getUserWithJpaSpecificationExecutor(LocalDateTime.of(2019, 1, 1, 0, 0, 0),
            LocalDateTime.of(2020, 1, 1, 0, 0, 0),
            "陈");
    assert queryWithTimeAndKeyWord != null;
    //根据时间查询
    List<User> queryWithTime =
getUserWithJpaSpecificationExecutor(LocalDateTime.of(2019, 1, 1, 0, 0, 0),
            LocalDateTime.of(2020, 1, 1, 0, 0, 0),
            null);
    assert queryWithTime != null;
    //普通查询
    List<User> query = getUserWithJpaSpecificationExecutor(null, null, "张");
    assert query != null;
}

private List<User> getUserWithJpaSpecificationExecutor(@Nullable
LocalDateTime start,
    @Nullable LocalDateTime end, @Nullable String keyWord) {
        //String.format 中%为特殊字符需要再加一个%进行转义，以下格式化的结果为%{keyWord}%
        String nameLike = keyWord == null ? null : String.format("%%%s%%", keyWord);
        return userRepository.findAll(((root, query, criteriaBuilder) -> {
            //根据传入参数的不同构造谓词列表
            List<Predicate> predicates = new ArrayList<>();
            if (start != null && end != null) {
                predicates.add(criteriaBuilder.between(root.get("createTime"),
start, end));
            }
            if (!StringUtils.isEmpty(nameLike)) {
                predicates.add(criteriaBuilder.like(root.get("name"), nameLike));
            }
            query.where(predicates.toArray(new Predicate[0]));
            return query.getRestriction();
        }));
}
```

较之于 JpaRepository 的查询方式，JpaSpecificationExecutor 并不需要另外定义接口，通过组合各种谓词（Predicate）构造最终的查询条件。

4.2.5 基于 JPQL 或 SQL

为了适应需要用到 SQL 语句的场景，JPA 同样支持基于 SQL 语句的查询，并且考虑到不同数据库的 SQL 语法存在差异，另外提供独立于平台的 JPQL（Java Persistence Query Language）。

【示例 4-3】

该类方式依赖注解@Query 实现。示例代码如下：

```
//JPQL
@Query("from User user0_ where (user0_.createTime between ?1 and ?2) and (user0_.name like ?3)")
List<User> query(LocalDateTime start, LocalDateTime end, String keyWord);

//SQL
@Query(nativeQuery = true,
    value = "select user0_.id as id1_1_, " +
        "user0_.account as account2_1_, " +
        "user0_.create_time as create_t3_1_, " +
        "user0_.name as name4_1_, " +
        "user0_.password as password5_1_, " +
        "user0_.vehicle_id as vehicle_6_1_ " +
        "from user user0_ " +
        "where (user0_.create_time between ? and ?) and (user0_.name like ?)")
List<User> queryNative(LocalDateTime start, LocalDateTime end, String keyWord);
```

JPQL 与 SQL 通过 nativeQuery 参数进行区分，当 nativeQuery 为 True 时，查询语句将被当作 SQL 处理。JPQL 语法与 SQL 非常相似，区别在于 JPQL 对表与字段的描述是使用对应的实体类与属性进行表达。

例如，以上示例中的 from User user0_ where (user0_.createTime between ?1 and ?2) and (user0_.name like ?3)，其中 User 代表 User 实体类的类名，createTime 与 name 为 User 实体类当中的属性名。这些元素在 JPQL 都是区分字母大小写的，需要与实体类中的定义保持一致。这意味着当该 JPQL 被改写为 from user user0_ where (user0_.create_time between ?1 and ?2) and (user0_.name like ?3)时，将会引起报错。

4.2.6 多表连接

Spring Data JPA 在处理复杂业务系统的优势之一，在于能够像对待对象一样管理两个表之间的关系。这使得在使用对象模型映射数据库属性的时候更加方便。根据业务逻辑和建模方式，可以创建单向或双向的关系。

表与表之间的关系可以通过如下注解表达并处理：

- @OneToOne：一对一（双向）。

- @OneToMany：一对多（双向）。
- @ManyToOne：多对一（单向）。
- @ManyToMany：多对多（双向）。

@OneToOne

假设目前有两张表 class 与 class_room，两张表的关系为一对一，其中 class 表包含外键 room_id，如图 4.4 所示。

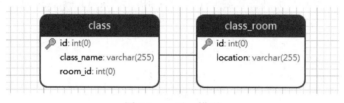

图 4.4　一对一模型

使用 @OneToOne 声明表关系时，首先需要关注的是外键的所有者。通常情况下，在外键所有者的实体类中，@OneToOne 注解需要配合 @JoinColumn 注解使用。@JoinColumn 在其中用于声明外键字段名。

【示例 4-4】

示例代码 Class.java：

```java
@Entity
@Getter
@Setter
@Accessors(chain = true)
@Table(name = "class")
public class Class implements Serializable {

    private static final long serialVersionUID = 1L;

    @Id
    @Column(name = "id", nullable = false)
    private Integer id;
    @Column(name = "class_name")
    private String className;

    @OneToMany(fetch = FetchType.LAZY, mappedBy = "clazz")
    private List<Student> students;

    @OneToOne(fetch = FetchType.LAZY)
    @JoinColumn(name = "room_id")
    private ClassRoom classRoom;

}
```

如果需要通过 ClassRoom 访问对应的 Class 实体的内容，则需要创建双向关系。另外创建外键重复上述操作也是可行的，但这类操作在数据库设计层面显得多余。事实上 @OneToOne 提供了 mappedBy 属性用于该项需求。示例代码 ClassRoom.java：

```
@Entity
@Getter
@Setter
@Accessors(chain = true)
@Table(name = "class_room")
public class ClassRoom implements Serializable {

    private static final long serialVersionUID = 1L;

    @Id
    @Column(name = "id", nullable = false)
    private Integer id;

    @Column(name = "location")
    private String location;

    @OneToOne(mappedBy = "classRoom", fetch = FetchType.LAZY)
    private Class clazz;

}
```

1. @OneToMany/@ManyToOne

比如表 Student 与表 Class，两表的关系为一对多/多对一，如图 4.5 所示。

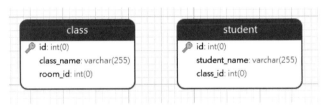

图 4.5　一对多/多对一模型

示例代码 Student.java：

```
@Entity
@Table(name = "student")
@Getter
@Setter
@Accessors(chain = true)
public class Student implements Serializable {

    private static final long serialVersionUID = 1L;

    @Id
    @Column(name = "id", nullable = false)
    private Integer id;

    @Column(name = "student_name")
    private String studentName;

    @ManyToOne(fetch = FetchType.LAZY)
    @JoinColumn(name = "class_id")
    private Class clazz;
```

}

在一对多/多对一的逻辑关系下，一般为多的实体类为外键的所有者，故同样需要使用@JoinColumn 配合使用，用以声明外键字段名，并且因为多对一为单向关系的缘故，不能使用 mappingBy 属性。

2. @ManyToMany

比如表 Teacher 和表 Class，两表关系为多对多。两表通过中间表 teacher_class 建立逻辑关系，如图 4.6 所示。

图 4.6　多对多模型

多对多的逻辑关系相对特殊，因为有中间表的存在，需要借助@JoinTable 描述中间表中包含的连接字段。示例代码 Teacher.java：

```
@Entity
@Table(name = "teacher")
@Getter
@Setter
@Accessors(chain = true)
public class Teacher implements Serializable {

    private static final long serialVersionUID = 1L;

    @Id
    @Column(name = "id", nullable = false)
    private Integer id;

    @Column(name = "teacher_name")
    private String teacherName;

    @ManyToMany
    @JoinTable(name = "teacher_class",
        joinColumns = {@JoinColumn(name = "class_id")},
        inverseJoinColumns = {@JoinColumn(name = "teacher_id")})
    private Set<Class> classes;

}
```

其中需要使用@JoinTable 注解中的 name 属性声明中间表的表名，joinColumns 声明正向连接的字段名，还需要 inverseJoinColumns 声明反向连接的字段名。

> **提　示**
>
> 因为@ManyToMany 是双向关系，故另一表内可以免去@JoinTable 这一批注，改用 mappingBy 属性。

4.2.7 级联操作

级联操作基于多表连接，在诸如@OneToOne、@OneToMany 这些关系注解中，包含一个 cascade 属性用于设置表之间的级联操作，其作用在于描述多个表在更新操作过后所发生的级联反应。

级联操作分不同等级，各等级如下：

- PERSIST：级联保存。当前实体如果被保存，则与之连接的实体也将会被保存。
- REMOVE：级联删除。当前实体如果被删除，则与之连接的实体也将会被删除。
- MERGE：级联合并（更新）。当前实体中的数据如果改变，将会相应地改变与之连接的实体。
- REFRESH：级联刷新。假设有一个订单，订单里关联若干商品。该订单可以被很多人操作。如果用户 A 对此订单和关联的商品进行了修改，同时 B 也进行了相同的操作，但 B 更先一步比 A 保存了数据。那么当 A 保存数据的时候，就需要先刷新订单信息及关联的商品信息，再将订单及商品保存。
- DETACH：级联脱离。将实体与其他实体的级联关系进行分离。
- ALL：包含以上所有权限。

4.2.8 加载类型

加载类型（FetchType）基于多表连接，同样是各关系注解中的配置项之一。加载类型分为 EAGER 与 LAZY 两种类型，其中 LAZY 为默认值。当加载类型为 EAGER 时，相关联的实体类将会立刻加载，即应用程序将立刻执行 SQL 语句获取数据。而 LAZY 类型则恰恰相反，相关联实体在需要时才会执行 SQL 语句。

示例代码：

```
@Test
public void testOneToOne() {
    Class clazz = classRepository.getOne(1);
    ClassRoom room = clazz.getClassRoom();
    assert room != null;
    assert room.getLocation() != null;
}
```

当 FetchType 为 LAZY 时，输出 SQL 日志如下：

```
Hibernate :
SELECT
class0_.id AS id1_0_0_,
class0_.class_name AS class_na2_0_0_,
class0_.room_id AS room_id3_0_0_
FROM
 class class0_
WHERE
 class0_.id =?
Hibernate : SELECT
classroom0_.id AS id1_1_0_,
classroom0_.location AS location2_1_0_
```

```
FROM
 class_room classroom0_
WHERE
 classroom0_.id =?
```

当 FetchType 为 EAGER 时，输出 SQL 日志如下：

```
Hibernate :
SELECT
class0_.id AS id1_0_0_,
class0_.class_name AS class_na2_0_0_,
class0_.room_id AS room_id3_0_0_,
classroom1_.id AS id1_1_1_,
classroom1_.location AS location2_1_1_
FROM
 class class0_
 LEFT OUTER JOIN class_room classroom1_ ON class0_.room_id = classroom1_.id
WHERE
 class0_.id =?
```

4.3 Spring Data MongoDB

除了传统的关系型数据库之外，在当前流行的应用程序开发中，非关系型数据库也占有不小的比重。在众多非关系型数据库中，MongoDB 是最为流行的非关系型数据库之一。MongoDB 定义为文档型数据库，它具有高性能、易部署并且易使用的特点。本节将介绍如何借助 Spring Data MongoDB 集成 MongoDB。

4.3.1 准备工作

Spring Data MongoDB 的配置非常简单，依赖方面引入 Spring Data 提供的 Starter 即可：

```xml
<dependency>
    <groupId>org.springframework.boot</groupId>
    <artifactId>spring-boot-starter-data-mongodb</artifactId>
</dependency>
```

Spring Data MongoDB 提供的数据访问大致基于 MongoTemplate 与 MongoRepository 这两种方式。MongoTemplate 遵循 Spring Boot 的标注模板形式，是在官方客户端的基础之上封装的持久化引擎。MongoRepository 则是按照 Spring Data 这个"大家族"中通用的设计模式所设计的 API。

Spring Data MongoDB 可以通过创建配置类继承 AbstractMongoClientConfiguration 进行配置，或者通过声明 MongoClient 与 MongoTemplate 的 JavaBean 实现。

【示例 4-5】

继承 AbstractMongoClientConfiguration 的示例代码：

```java
@Configuration
public class MongoConfig extends AbstractMongoClientConfiguration {
```

```
    @Override
    protected String getDatabaseName() {
        return "test";
    }

    @Override
    public MongoClient mongoClient() {
        ConnectionString connectionString = new
ConnectionString("mongodb://localhost:27017/test");
        MongoClientSettings mongoClientSettings =
MongoClientSettings.builder()
            .applyConnectionString(connectionString)
            .build();

        return MongoClients.create(mongoClientSettings);
    }

    @Override
    public Collection getMappingBasePackages() {
        return Collections.singleton("com.example");
    }
}
```

返回 JavaBean 的示例代码：

```
@Configuration
public class MongoConfig {

    @Bean
    public MongoClient mongo() {
        ConnectionString connectionString = new
ConnectionString("mongodb://localhost:27017/test");
        MongoClientSettings mongoClientSettings =
MongoClientSettings.builder()
            .applyConnectionString(connectionString)
            .build();

        return MongoClients.create(mongoClientSettings);
    }

    @Bean
    public MongoTemplate mongoTemplate() throws Exception {
        return new MongoTemplate(mongo(), "test");
    }
}
```

如果需要启用 Repository 对数据库进行操作，则需要加上注解@EnableMongoRepositories (basePackages = "com.example.mongodb.repository")。另外还需要创建对应的 Repository 接口，示例代码如下：

```
public interface WareRepository extends MongoRepository<Ware, String> {
    //Ware 为实体类型，String 为实体的主键类型
}
```

4.3.2 使用 MongoTemplate 访问 MongDB

Spring Data Mongodb 中基本文档查询依赖于 MongoTemplate。本小节将介绍基于 MongoTemplate 基础的增删改查。

1. <T> T insert(T objectToSave, String collectionName);

该方法用于在初始化时将数据写入到数据库当中。例如数据库内容为空。

执行代码：

```
Ware mineralWater = new Ware();
mineralWater.setName("矿泉水");
mongoTemplate.insert(mineralWater, "ware");
```

执行结果：

```
{
    "_id" : ObjectId("5f5c76db61083f0fb8ce7581"),
    "name" : "矿泉水",
    "_class" : "com.example.mongodb.domain.Ware"
}
```

2. <T> T save(T objectToSave);

该方法用于保存一个实体。当实体不存在数据库中时，保存操作会被视为"插入"。当实体已存在于数据库中（通过 Id 判定实体是否存在于数据库中），保存操作将被视为"更新"。例如数据库中已存在一个 Id 为 1 的实体记录：

```
{
    "_id" : "1",
    "_class" : "com.example.mongodb.domain.Ware"
}
```

执行代码：

```
Ware melonSeeds = new Ware();
melonSeeds
        .setId("1")
        .setName("瓜子");
mongoTemplate.save(melonSeeds);
```

执行结果：

```
{
    "_id" : "1",
    "name" : "瓜子",
    "_class" : "com.example.mongodb.domain.Ware"
}
```

3. UpdateResult updateFirst(Query query, UpdateDefinition update, Class<?> entityClass);

updateFirst 方法用于更新通过 Query 查询到的第一条实体记录。例如数据库中存在名称相同的多条记录：

```
[
    {
        "_id":ObjectId("5f5c706f81ffe73b0b5762e4"),
        "name":"Milk",
        "_class":"com.example.mongodb.domain.Ware"
    },{
        "_id":ObjectId("5f5c70dd8da6441372dfc517"),
        "name":"Milk",
        "_class":"com.example.mongodb.domain.Ware"
    }
]
```

执行代码：

```
Query query = new Query();
query.addCriteria(Criteria.where("name").is("Milk"));
Update update = new Update();
update.set("name", "Yogurt");
mongoTemplate.updateFirst(query, update, Ware.class);
```

执行结果：

```
[
    {
        "_id":ObjectId("5f5c706f81ffe73b0b5762e4"),
        "name":"Yogurt",
        "_class":"com.example.mongodb.domain.Ware"
    },{
        "_id":ObjectId("5f5c70dd8da6441372dfc517"),
        "name":"Milk",
        "_class":"com.example.mongodb.domain.Ware"
    }
]
```

4. UpdateResult updateMulti(Query query, UpdateDefinition update, Class<?> entityClass);

updateMulti 方法根据 Query 对象查询到多条数据，对该批数据进行批量更新。例如数据库中有多条 name 相同的实体对象：

```
[
{
        "_id" : ObjectId("5f5c7d7993b72416be19c44e"),
        "name" : "矿泉水",
        "status" : "selling",
        "_class" : "com.example.mongodb.domain.Ware"
    },{
        "_id" : ObjectId("5f5c7d7eda0db17b306fad4c"),
        "name" : "矿泉水",
        "status" : "selling",
        "_class" : "com.example.mongodb.domain.Ware"
    }
]
```

执行代码：

```
Query query = new Query();
query.addCriteria(Criteria.where("name").is("矿泉水"));
Update update = new Update();
update.set("status", "sold out");
mongoTemplate.updateMulti(query, update, Ware.class);
```

执行结果：

```
[
{
        "_id" : ObjectId("5f5c7d7993b72416be19c44e"),
        "name" : "矿泉水",
        "status" : "sold out",
        "_class" : "com.example.mongodb.domain.Ware"
    },{
        "_id" : ObjectId("5f5c7d7eda0db17b306fad4c"),
        "name" : "矿泉水",
        "status" : "sold out",
        "_class" : "com.example.mongodb.domain.Ware"
    }
]
```

5. <T> T findAndModify(Query query, UpdateDefinition update, Class<T> entityClass);

findAndModify 方法根据 Query 查询对应实体对其进行更新，类似于 updateFirst。只不过方法的返回值为更新前的记录，例如数据库中有实体如下：

```
{
    "_id" : ObjectId("5f5c76db61083f0fb8ce7581"),
    "name" : "矿泉水",
    "_class" : "com.example.mongodb.domain.Ware",
    "status" : "sold out"
}
```

执行代码：

```
Query query = new Query();
query.addCriteria(Criteria.where("name").is("矿泉水"));
Update update = new Update();
update.set("status", "selling");
Ware result = mongoTemplate.findAndModify(query, update, Ware.class);
assert Optional.ofNullable(result).orElse(new
Ware()).getStatus().equals("sold out");
```

执行结果：

```
{
    "_id" : ObjectId("5f5c76db61083f0fb8ce7581"),
    "name" : "矿泉水",
    "_class" : "com.example.mongodb.domain.Ware",
    "status" : "selling"
}
```

6. UpdateResult upsert(Query query, UpdateDefinition update, Class<?> entityClass);

upsert 方法用于查找并更新或者创建实体。类似于 save 方法，当根据 Query 查找实体失败时，则创建对应实体。区别在于该方法根据 Query 判断实体是否存在，而 save 根据 Id 进行判断。例如数据库中无内容。

执行代码：

```
Query query = new Query();
query.addCriteria(Criteria.where("name").is("西瓜"));
Update update = new Update();
update.set("name", "哈密瓜");
mongoTemplate.upsert(query, update, Ware.class);
```

执行结果：

```
{
    "_id" : ObjectId("5f5caf71d28dbdbeb1aa261a"),
    "name" : "哈密瓜"
}
```

如果数据库中内容为：

```
{
    "_id" : ObjectId("5f5cf4120fb5c25ea0de99ba"),
    "name" : "西瓜",
    "status" : "selling",
    "_class" : "com.example.mongodb.domain.Ware"
}
```

执行结果：

```
{
    "_id" : ObjectId("5f5cf4120fb5c25ea0de99ba"),
    "name" : "哈密瓜",
    "status" : "selling",
    "_class" : "com.example.mongodb.domain.Ware"
}
```

7. DeleteResult remove(Query query, Class<?> entityClass);

remove 方法用于根据 Query 查找对应实体对象，并将符合条件的对象从数据库中移除。例如数据库中 status 为 sold out 的对象如下：

```
[
    {
        "_id" : ObjectId("5f5c7d7993b72416be19c44e"),
        "name" : "矿泉水",
        "status" : "sold out",
        "_class" : "com.example.mongodb.domain.Ware"
    },{
        "_id" : ObjectId("5f5c7d7eda0db17b306fad4c"),
        "name" : "矿泉水",
        "status" : "sold out",
        "_class" : "com.example.mongodb.domain.Ware"
    }
]
```

执行代码:

```
mongoTemplate.remove(Query.query(Criteria.where("status").is("sold out")),
Ware.class);
```

最终实体对象被移除。

4.3.3 使用 MongoRepository 访问 MongoDB

MongoRepository 与 JPA 中的 Repository 很相似，它继承了 PagingAndSortingRepository<T, ID>、QueryByExampleExecutor<T>这两个接口。除提供基础的增删改查功能之外还提供了诸如排序、分页之类的功能，另外提供了通过 Example 这一匹配对象的方式。

- <S extends T> Iterable<S> saveAll(Iterable<S> entities);

saveAll 方法用于保存所有给出的实体对象。示例代码如下:

```
List<Ware> wares = new ArrayList<>();
wares.add(new Ware().setName("酒水").setStatus("selling"));
wares.add(new Ware().setName("瓜子").setStatus("selling"));
wares.add(new Ware().setName("饮料").setStatus("selling"));
wareRepository.saveAll(wares);
```

- <S extends T> boolean exists(Example<S> example);

根据 Example 判断是否存在与 example 匹配的元素。例如数据库中不含 status 为 sold out 的元素，仅包含 status 为 selling 的元素。示例代码如下:

```
boolean isSoldOutExist = wareRepository.exists(Example.of(new
Ware().setStatus("sold out")));
   assert !isSoldOutExist;
   boolean isSellingExist = wareRepository.exists(Example.of(new
Ware().setStatus("selling")));
   assert isSellingExist;
```

- Iterable<T> findAll(Sort sort);

该方法将会查出所有记录，结果按排序规则进行排序。示例代码如下:

```
List<Ware> wareList = wareRepository.findAll(Sort.by(Sort.Direction.ASC,
"id"));
```

- Page<T> findAll(Pageable pageable);

该方法将会查出所有记录，结果按分页规则进行分页。示例代码如下:

```
Page<Ware> warePage = wareRepository.findAll(PageRequest.of(0, 10));
```

4.4　Spring Data Redis

在 NoSQL 数据库这个分类中，Redis 也是非常流行并且极具特色的一款产品。与适合处理海量数据易于扩展的 MongoDB 不同，Redis 在性能方面更为突出。Redis 作为内存型的键值数据库，以高度优化的存储结构将数据保存至内存，这使得该产品的性能可以在各数据库之中排到第一梯队。也是因为性能如此突出，在大多应用开发中，Redis 往往会被当成缓存使用，用于缓存其他数据库中的热点数据，以提高整个系统的查询性能。

本节将使用 Spring Data Redis 来整合 Redis 数据库。Spring Data Redis 与之前所介绍的 Spring Data 相关项目类似，会为开发人员提供模板以及 Repository 这两种通用的数据访问形式。

4.4.1　准备工作

为了能够使用到 Spring Data Redis 项目所提供的功能，需要引入对应的 starter 依赖：

```xml
<dependency>
    <groupId>org.springframework.boot</groupId>
    <artifactId>spring-boot-starter-data-redis</artifactId>
</dependency>
```

另外，连接 Redis 有两种客户端解决方案：Lettuce 与 Jedis。Spring Data Redis 默认集成了 Lettuce 作为连接客户端。如果要另外使用 Jedis，需要另外引入对应的依赖。

```xml
<dependency>
    <groupId>redis.clients</groupId>
    <artifactId>jedis</artifactId>
    <version>3.3.0</version>
</dependency>
```

在引入依赖之后，需要声明一个配置项用于启用 Repository 以及模板。

示例代码如下：

```
@Bean
public LettuceConnectionFactory redisConnectionFactory() {
    //使用 Lettuce 作为客户端需要声明该 Bean
    RedisStandaloneConfiguration redisStandaloneConfiguration =
new RedisStandaloneConfiguration(hostName, port);
    //如果有密码，则需要通过 setPassword 设置对应密码
    redisStandaloneConfiguration.setPassword(password);
    return new LettuceConnectionFactory(redisStandaloneConfiguration);
}

@Bean
public JedisConnectionFactory jedisConnectionFactory() {
    //使用 Jedis 作为客户端需要声明该 Bean
    RedisStandaloneConfiguration redisStandaloneConfiguration =
new RedisStandaloneConfiguration(hostName, port);
    redisStandaloneConfiguration.setPassword(password);
```

```java
        return new JedisConnectionFactory(redisStandaloneConfiguration);
    }

    @Bean
    public RedisTemplate<?, ?> redisTemplate() {
        RedisTemplate<String, Object> template = new RedisTemplate<>();
        RedisSerializer<String> stringSerializer = new StringRedisSerializer();
        JdkSerializationRedisSerializer jdkSerializationRedisSerializer = new JdkSerializationRedisSerializer();
        template.setConnectionFactory(redisConnectionFactory());
        template.setKeySerializer(stringSerializer);
        template.setHashKeySerializer(stringSerializer);
        template.setValueSerializer(jdkSerializationRedisSerializer);
        template.setHashValueSerializer(jdkSerializationRedisSerializer);
        template.setEnableTransactionSupport(true);
        template.afterPropertiesSet();
        return template;
    }
```

其中 LettuceConnectionFactory 与 JedisConnectionFactory 根据所选的客户端选择声明对应 ConnectionFactory 即可。

4.4.2 使用 RedisRepository 访问 Redis

RedisRepository 延续了 Spring Data 中通用的 Repository 的设计风格，为操作 Redis 带来了简单易用的方式。

【示例 4-6】

实体示例代码：

```java
@Accessors(chain = true)
@Getter
@Setter
@RedisHash(value = "Student",timeToLive = 10)
public class Student implements Serializable {

    public enum Gender {
        MALE, FEMALE
    }

    private String id;
    @Indexed
    private String name;
    private Gender gender;
    private int grade;
}
```

其中@RedisHash 注解用于声明该实体将被存储于 Redis Hash 中，如果需要使用 Repository 的数据访问形式，这个注解是必须使用到的。其中 timeToLive 属性用于标注该实体对象在数据库中的有效期，单位为秒。@Indexed 注解用于标注需要作为查询条件的属性。

示例代码 StudentRepository.java：

```java
@Repository
public interface StudentRepository extends CrudRepository<Student, String> {

    //自定义查询方式。使用标注了@Indexed 的 name 属性进行查询
    Student findByName(String name);

}
```

调用示例代码如下:

```java
@SpringBootTest
class RedisApplicationTests {

    @Autowired
    StudentRepository studentRepository;

    @Test
    void testSave() {
        Student student = new Student()
                .setId("20200101007")
                .setName("zbc")
                .setGender(Student.Gender.MALE)
                .setGrade(1);
        //根据 Id 新增或更新记录
        studentRepository.save(student);
    }

    @Test
    void testFindBy() {
        //使用主键查询
        assert studentRepository.findById("20200101007").isPresent();
        //根据自定义方法查询
        assert studentRepository.findByName("zbc") != null;
    }

    @Test
    void testDelete() {
        //根据主键删除
        studentRepository.deleteById("20200101007");
        assert !studentRepository.findById("20200101007").isPresent();
    }

    @Test
    void testFindAll() {
        studentRepository.save(new Student()
                .setId("20200101007")
                .setName("zbc")
                .setGender(Student.Gender.MALE)
                .setGrade(1));
        studentRepository.save(new Student()
                .setId("20200101008")
                .setName("cc")
                .setGender(Student.Gender.FEMALE)
                .setGrade(1));
        studentRepository.save(new Student()
```

```
            .setId("20200101009")
            .setName("gf")
            .setGender(Student.Gender.MALE)
            .setGrade(1));
        studentRepository.save(new Student()
            .setId("20200101010")
            .setName("mjyl")
            .setGender(Student.Gender.FEMALE)
            .setGrade(1));
        //查询全部记录
        List<Student> studentList =
Lists.newArrayList(studentRepository.findAll());
        assert studentList.size() > 0;
    }
}
```

4.4.3 使用 RedisTemplate 访问 Redis

相对于 RedisRepository，RedisTemplate 更为灵活。RedisTemplate 基于 Redis 的原生命令封装了一系列的操作方法。这些方法基于 Redis 数据结构实现，Redis 的基础数据结构有 5 种：

- String：字符串，Redis 中基础的数据结构。
- List：列表，该数据结构底层的存储形式为链表。插入与删除的操作效率非常高。
- Hash：哈希字典，与 Java 中的 HashMap 非常类似。底层结构为数组与链表。
- Set：集合，与 Java 中的 HashSet 类似。内部的键值对是无序且唯一的。
- Zset：有序集合，又称为 SortedSet。它一方面保证键值对唯一，另一方面会维护键值对的权重属性，为键值对进行排序。

【示例 4-7】

操作 RedisTemplate 的示例代码如下：

```
@SpringBootTest
public class RedisTemplateTests {

    @Autowired
    RedisTemplate<String, Object> redisTemplate;


    @Test
    void testString() {
        //设置键-值对
        redisTemplate.opsForValue().set("num", "123");
        //根据 name 获取值
        redisTemplate.opsForValue().get("num");
        //设置带有效期的键-值对
        redisTemplate.opsForValue().set("fade-num", "321", 10, TimeUnit.SECONDS);
        //10 秒后结果为 Null
        redisTemplate.opsForValue().get("fade-num");
    }
```

```java
    @Test
    void testList() {
        //通过leftPush更新列表，也可以选择rightPush方法
        redisTemplate.opsForList().leftPush("languages", "java");
        redisTemplate.opsForList().leftPush("languages", "python");
        redisTemplate.opsForList().leftPush("languages", "c++");
        //查询列表长度
        assert redisTemplate.opsForList().size("languages") >= 3;
        //弹出列表左边的元素，之后结果在数据库中不复存在
        assert Objects.equals(redisTemplate.opsForList().leftPop("languages"), "c++");
        //弹出列表右边的元素，之后结果在数据库中不复存在
        assert Objects.equals(redisTemplate.opsForList().rightPop("languages"), "java");
    }

    @Test
    void testHash() {
        //更新hash 第一个参数为hash的键，第二个参数为该hash内键-值对的键
        redisTemplate.opsForHash().put("hash", "red", "小红");
        redisTemplate.opsForHash().put("hash", "elephant", "小象");
        redisTemplate.opsForHash().put("hash", "red-elephant", "小红象");
        //根据hash的键以及hash内键值对的键检索对应信息
        assert Objects.equals(redisTemplate.opsForHash().get("hash", "red"), "小红");
        //获取hash内所有键-值对的键
        Set<Object> hash = redisTemplate.opsForHash().keys("hash");
        hash.forEach(System.out::println);
    }

    @Test
    void testSet() {
        //新增set
        redisTemplate.opsForSet().add("set", "sir", "yes", "sir", "madam");
        //随机弹出一个元素
        redisTemplate.opsForSet().pop("set");
        //获取set中所有内容
        redisTemplate.opsForSet().members("set").forEach(System.out::println);
    }

    @Test
    void testZSet() {
        //新增zset内容
        redisTemplate.opsForZSet().add("zset", "sir", -1);
        redisTemplate.opsForZSet().add("zset", "sir", 9);
        redisTemplate.opsForZSet().add("zset", "yes", 3);
        redisTemplate.opsForZSet().add("zset", "madam", 10);
        //获取该zset中sir的权重
        System.out.println(redisTemplate.opsForZSet().score("zset", "sir"));
        //根据权重区间遍历zset
        redisTemplate.opsForZSet().range("zset", 0, 9).forEach(System.out::println);
    }

}
```

第 5 章

应用程序安全性

Web 应用的开发中不乏对程序安全性的需求。当下流行的各种社交软件，一般需要通过登录验证保护用户的个人资源，其中的会员制度将普通用户享用的功能与付费用户享用的功能加以区分。各种类型的管理系统更是需要根据角色与权限实现复杂的控制。为了帮助开发人员实现这些需求，Spring 社区创建了 Spring Security 这一项目。该项目几乎覆盖了认证与授权相关的方方面面。

本章将介绍不同场景下 Spring Boot 整合 Spring Security 的示例，涉及的知识点有：

- 基础注册登录流程
- 权限管理
- 基于 Session Cookie 的验证
- 基于 Token 的验证
- OAuth 2.0 授权

5.1 基于 Spring Security 的注册登录

Spring Security 的前身是 Acegi Security，之后作为 Spring 中的子项目，经过完善更名为 Spring Security。作为一个安全框架，需要解决的问题大多可以总结为两个方面：认证（你是谁？）与授权（你被允许做什么？）。Spring Security 架构旨在分开解决这两方面，并且提供了不少策略以及扩展点。本节将通过介绍基础的注册登录示例，大致了解集成 Spring Security 的基本步骤。

5.1.1 Spring Security 简介

Spring Security 的整体架构可以概括为一组过滤器链。这组过滤器链在项目启动时进行自动配置，不同过滤器有不同职责，共同作用以实现应用程序的安全管理，如图 5.1 所示。

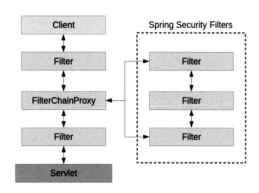

图 5.1　Spring Security 架构图

在实际的软件架构上来看，Spring Security 本质是一个过滤器。它本身不会做实质上的验证，但是会将请求委派给 Spring Security 的内部过滤器链进行进一步的处理。其内部过滤器链可以有多个，根据配置策略将请求区分开，并分配到不同的过滤器链上。常见的有基于匹配请求路径的调度方式，如图 5.2 所示。

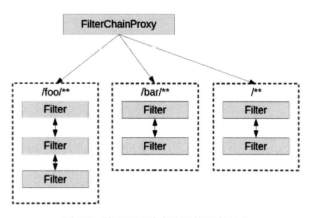

图 5.2　基于匹配请求路径的调度方式

5.1.2　用户注册

常见的 Web 应用大多基于关系型数据库做用户管理，故该示例将基于 Spring Data JPA 构建应用程序的数据访问层。

【示例 5-1】

用户实体类 TUser.java：

```
@Accessors(chain = true)
@Setter
@Getter
@Entity
@Table(name = "user")
```

```java
public class TUser {

    @GeneratedValue(strategy=GenerationType.IDENTITY)
    @Id
    @Column(name = "id", nullable = false)
    private Integer id;
    @Column(name = "username", nullable = false)
    private String username;
    @Column(name = "password", nullable = false)
    private String password;

}
```

UserRepository.java：

```java
public interface UserRepository extends JpaRepository<TUser, Integer>,
 JpaSpecificationExecutor<TUser> {

    TUser findByUsername(String username);

}
```

以上两个类构成用于实现用户管理的数据访问层。接下来，基于数据访问层实现用户注册相关的控制器层。控制器层内容包含用户注册以及用于表示受保护资源的部分。

统一响应体 R.java：

```java
@Accessors(chain = true)
@Setter
@Getter
public class R<T> {

    private Integer code;
    private T data;
    private String message;
    private static final Integer SUCCESS = 0;
    private static final Integer FAILED = -1;

    public static <T> R<T> ok(T data) {
        return new R<T>()
                .setCode(SUCCESS)
                .setData(data);
    }

    public static <T> R<T> failed(String message) {
        return new R<T>()
                .setCode(FAILED)
                .setMessage(message);
    }

}
```

用户 DTO 类 UserDto.java：

```java
@Accessors(chain = true)
@Setter
@Getter
```

```java
public class UserDto {

    @NotNull
    private String username;
    @NotNull
    private String password;

}
```

SimpleController.java：

```java
@RequiredArgsConstructor
@RestController
public class SimpleController {

    private final UserRepository userRepository;
    private final PasswordEncoder encoder;

    @PostMapping("/account")
    public R<Void> register(@RequestBody @Valid UserDto userDto) {
        TUser TUser = new TUser()
                .setUsername(userDto.getUsername())
                //数据库中存储明文密码会降低系统安全性，故存储前使用encoder为密码加密
                .setPassword(encoder.encode(userDto.getPassword()));
        userRepository.save(TUser);
        return R.ok(null);
    }

    @GetMapping("/hello")
    public R<String> greeting() {
        //用来代表受保护的资源
        Authentication authentication =
SecurityContextHolder.getContext().getAuthentication();
        //获取当前用户名
        return R.ok("Hi! This is " + authentication.getName());
    }
}
```

5.1.3 用户登录

用户登录模块的实现依赖 Spring Security 提供的 formLogin 模式。formLogin 将表单登录所需的绝大多数逻辑在框架内实现了，另外开放了接口与配置供开发人员自定义登录验证相关的业务逻辑。UserDetailService 是其中的关键一环，在开发过程中实现 UserDetailsService 用于表示用户信息被载入的方式。

【示例 5-2】

UserDetailsService 实现类 SimpleUserDetailServiceImpl.java：

```java
@Service
@RequiredArgsConstructor
public class SimpleUserDetailServiceImpl implements UserDetailsService {
```

```java
    private final UserRepository userRepository;

    @Override
    public UserDetails loadUserByUsername(String username) throws UsernameNotFoundException {
        TUser tUser = userRepository.findByUsername(username);
        if (tUser == null) {
            throw new UsernameNotFoundException("username not found");
        }
        return new User(tUser.getUsername(), tUser.getPassword(), getAuthorities());
    }

    private Collection<GrantedAuthority> getAuthorities() {
        //获取用户的角色权限，本示例不包含权限控制，故简化实现过程
        return Collections.singletonList(new SimpleGrantedAuthority("ROLE_USER"));
    }
}
```

因为 Spring Security 已经将登录验证相关的逻辑在框架中实现了，剩下的工作仅仅通过编写配置即可完成。Spring Security 配置类 SecurityConfig.java：

```java
@Configuration
@RequiredArgsConstructor
public class SecurityConfig extends WebSecurityConfigurerAdapter {

    private final ObjectMapper mapper;
    private final DataSource dataSource;
    private final UserDetailsService userDetailsService;

    @Bean
    public PasswordEncoder bCryptPasswordEncoder() {
        return new BCryptPasswordEncoder();
    }

    @Override
    protected void configure(HttpSecurity http) throws Exception {
        http
                //开启登录配置
                .authorizeRequests()
                //表示 account 接口的访问不受限制
                .antMatchers("/account").permitAll()
                //表示所有接口，登录之后就能访问
                .anyRequest().authenticated()
                .and()
                .formLogin()
                //.loginPage("/login") 添加该项将定义登录页面，未登录时，访问一个需要登录之后才能访问的接口，就会自动跳转到该页面
                //登录处理接口
                .loginProcessingUrl("/session")
                //定义登录时，用户名的 key，默认为 username
                .usernameParameter("uname")
                //定义登录时，用户密码的 key，默认为 password
                .passwordParameter("pwd")
```

```
            //登录成功的处理器
            .successHandler((req, resp, authentication) -> {
                resp.setContentType("application/json;charset=utf-8");
                PrintWriter out = resp.getWriter();
                out.write(mapper.writeValueAsString(R.<Void>ok(null)));
                out.flush();
            })
            //登录失败的处理器
            .failureHandler((req, resp, exception) -> {
                resp.setContentType("application/json;charset=utf-8");
                PrintWriter out = resp.getWriter();
                out.write(mapper.writeValueAsString(R.<Void>failed("failed")));
                out.flush();
            })
            //和表单登录相关的接口可以不受限制直接通过
            .permitAll()
            .and()
            .logout()
            .logoutUrl("/logout")
            //登出成功的处理程序
            .logoutSuccessHandler((req, resp, authentication) -> {
                resp.setContentType("application/json;charset=utf-8");
                PrintWriter out = resp.getWriter();
                out.write(mapper.writeValueAsString(R.<Void>ok(null)));
                out.flush();
            })
            .permitAll()
            .and()
            .httpBasic()
            .disable()
            .csrf()
            .disable();
    }
}
```

配置中可以通过 loginPage 方法自定义登录页的路径，当处于未登录状态并且先访问网页资源时，框架将会自动跳转至所配置的路径。如果不通过该方法做自定义配置，框架将自动生成一个登录页。

5.1.4 "记住我"功能

formLogin 认证模式提供了 "Remember Me"（记住我）相关的选项。通过配置相关选项，客户端在登录成功后会获得 "记住我" 对应的 token 凭证。在有效期内关闭浏览器，重新访问受保护的资源都享有免登录的权利。实现方式如下：

首先，在数据库内创建一张用于记录 token 的表 persistent_logins。

【示例 5-3】

persistence_logins.sql：

```sql
CREATE TABLE persistent_logins (
  username VARCHAR ( 64 ) NOT NULL,
  series VARCHAR ( 64 ) PRIMARY KEY,
  token VARCHAR ( 64 ) NOT NULL,
  last_used TIMESTAMP NOT NULL)
```

接下来，修改 SecurityConfig.java 的内容：

```java
@Configuration
@RequiredArgsConstructor
public class SecurityConfig extends WebSecurityConfigurerAdapter {

    private final ObjectMapper mapper;
    private final DataSource dataSource;
    private final UserDetailsService userDetailsService;

    @Override
    protected void configure(HttpSecurity http) throws Exception {
        //……省略若干代码
        //登出成功的处理程序
        .logoutSuccessHandler((req, resp, authentication) -> {
            resp.setContentType("application/json;charset=utf-8");
            PrintWriter out = resp.getWriter();
            out.write(mapper.writeValueAsString(R.<Void>ok(null)));
            out.flush();
        })
        .permitAll()
        .and()
        //省略以上若干代码，以下加粗部分为新增内容
        .rememberMe()
        .rememberMeParameter("remember-me")
        .tokenRepository(persistentTokenRepository())
        .userDetailsService(userDetailsService)
        .and()
        .httpBasic()
        .disable()
        .csrf()
        .disable();
    }

    @Bean
    public PersistentTokenRepository persistentTokenRepository() {
        JdbcTokenRepositoryImpl tokenRepository = new JdbcTokenRepositoryImpl();
        tokenRepository.setDataSource(dataSource);
        return tokenRepository;
    }

}
```

5.2 权限管理

Spring Security 自带的权限管理可以在不同粒度中进行,分别是权限(Authority)和角色(Role)。这使得权限管理相关的业务逻辑开发可以非常灵活。本节将介绍基于 Spring Security 的权限管理模块的配置与开发流程。

5.2.1 权限与角色

"权限"指的是用户拥有对特定资源的某种特权。例如一些系统中存在一些重要数据可供用户进行读写操作,读写两种操作便可以被区分为读权限(READ_AUTHORITY)与写权限(WRITE_AUTHORITY)。被赋予读权限的才可以对资源进行读操作,被赋予写权限的才可以对资源进行写操作,两种权限互相独立互不影响。这使得权限管理可以用一种细粒度的方式进行管理。

相对粗粒度的权限管理方式便是"角色"。角色可以被视作一种"容器",容器的内容便是细粒度的"权限"。例如写权限被进一步拆分为新增(INSERT_AUTHORITY)、更新(UPDATE_AUTHORITY)与删除(DELETE_AUTHORITY)。角色分为游客(ROLE_ANONYMOUS)、普通管理员(ROLE_STAFF)与超级管理员(ROLE_ADMINISTRATOR)。游客仅具有读权限,普通管理员包含除删除权限外的其他权限,而超级管理员则拥有所有权限。这样可以将细粒度的权限进行聚合,为对应的业务开发提供便利。

5.2.2 权限管理体系中的实体:用户、角色与权限

权限管理体系中有三个主要的实体:用户、角色与权限。

【示例 5-4】

权限 TAuthority.java:

```java
@Accessors(chain = true)
@Setter
@Getter
@Entity
@Table(name = "authority")
public class TAuthority {

    @GeneratedValue(strategy = GenerationType.AUTO)
    @Id
    @Column(name = "id", nullable = false)
    private Integer id;
    @Column(name = "name", nullable = false)
    private String name;
    @ManyToMany(mappedBy = "authorities")
    private Collection<TRole> roles;
```

}

角色 TRole.java：

```java
@Accessors(chain = true)
@Setter
@Getter
@Entity
@Table(name = "role")
public class TRole {

    @GeneratedValue(strategy = GenerationType.AUTO)
    @Id
    @Column(name = "id", nullable = false)
    private Integer id;
    @Column(name = "name", nullable = false)
    private String name;

    @ManyToMany(mappedBy = "roles")
    private Collection<TUser> users;

    @ManyToMany
    @JoinTable(name = "roles_authorities", joinColumns = @JoinColumn(name = "role_id", referencedColumnName = "id"), inverseJoinColumns = @JoinColumn(name = "authority_id", referencedColumnName = "id"))
    private Collection<TAuthority> authorities;

}
```

用户 TUser.java：

```java
@Accessors(chain = true)
@Setter
@Getter
@Entity
@Table(name = "user")
public class TUser {

    @GeneratedValue(strategy = GenerationType.AUTO)
    @Id
    @Column(name = "id", nullable = false)
    private Integer id;
    @Column(name = "username", nullable = false)
    private String username;
    @Column(name = "password", nullable = false)
    private String password;

    @ManyToMany
    @JoinTable(
            name = "users_roles",
            joinColumns = @JoinColumn(name = "user_id", referencedColumnName = "id"), inverseJoinColumns = @JoinColumn(name = "role_id", referencedColumnName = "id"))
    private Collection<TRole> roles;

}
```

其中权限与角色以及角色与用户之间存在多对多的关系，因此在建表过程中还需要加入两张中间表。各表的模型图如图 5.3 所示。

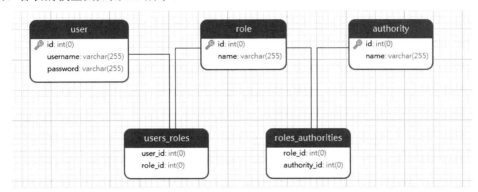

图 5.3　模型图

在该实体关系中，一个用户的权限与角色相关。UserDetailService 的实现类需要做相应的调整，调整后的实现类 AuthorityUserDetailServiceImpl.java：

```
@Transactional
@Service
@RequiredArgsConstructor
public class AuthorityUserDetailServiceImpl implements UserDetailsService {

    private final UserRepository userRepository;

    @Override
    public UserDetails loadUserByUsername(String username) throws UsernameNotFoundException {
        TUser tUser = userRepository.findByUsername(username);
        if (tUser == null) {
            throw new UsernameNotFoundException("username not found");
        }
        return new User(tUser.getUsername(), tUser.getPassword(), getAuthorities(tUser));
    }

    private Collection<GrantedAuthority> getAuthorities(TUser user) {
        List<GrantedAuthority> authorities = new ArrayList<>();
        //将用户的角色以及角色的权限一起加入至 GrantedAuthority 集合
        for (TRole role : user.getRoles()) {
            authorities.add(new SimpleGrantedAuthority(role.getName()));
            for (TAuthority authority : role.getAuthorities()) {
                authorities.add(new SimpleGrantedAuthority(authority.getName()));
            }
        }
        return authorities;
    }
}
```

5.2.3 配置与应用

权限管理最终还是需要进行配置才可以得以生效。

【示例 5-5】

受权限管理保护的控制器 Simp-leController.java：

```java
@RequiredArgsConstructor
@RestController
public class SimpleController {

    //……省略若干方法与属性

    @GetMapping("/administrator")
    public R<String> administrator() {
        //只有 administrator 才可以访问的接口
        return R.ok("administrator");
    }

    @GetMapping("/staff")
    public R<String> staff() {
        //只有 staff 才可以访问的接口
        return R.ok("staff");
    }

    @GetMapping("/read")
    public R<String> read() {
        //用以表示读操作接口
        return R.ok("blah blah blah……");
    }

    @PostMapping("/write")
    public R<String> write(@RequestParam String input) {
        //用以表示写操作接口
        return R.ok(input);
    }
}
```

使用 antMatchers 方法，通过路径匹配的方式实现最终的权限管理。SecurityConfig.java：

```java
@Configuration
@RequiredArgsConstructor
public class SecurityConfig extends WebSecurityConfigurerAdapter {

    @Override
    protected void configure(HttpSecurity http) throws Exception {
        http
                //开启登录配置
                .authorizeRequests()
                //表示 account 接口的访问不受限制
                .antMatchers("/account").permitAll()
                .antMatchers("/write").hasAuthority("WRITE_AUTHORITY")
```

```
                    .antMatchers("/read").hasAnyAuthority("READ_AUTHORITY",
"WRITE_AUTHORITY")
                    .antMatchers("/administrator").hasRole("ADMINISTRATOR")
                    .antMatchers("/staff").hasAnyRole("STAFF", "ADMINISTRATOR")
                    //表示所有接口，登录之后就能访问
                    .anyRequest().authenticated()
                    //……省略
            }
```

示例中 hasAuthority()、hasAnyAuthority()、hasRole()和 hasAnyRole()方法用以检查用户在访问对应路径时是否具有对应权限。

> **注 意**
>
> hasRole 和 hasAnyRole 会在检查流程中自动加入"ROLE_"前缀。例如对路径"/administrator"的检查是确认用户是否有名称为"ROLE_ADMINISTRATOR"的权限。

使用 antMatchers 进行路径匹配时，除了直接指定路径之外还可以基于 antPattern 写出灵活的匹配规则。antPattern 的规则如下：

- ?：匹配单个字符。
- *：匹配 0 个或者多个字符。
- **：匹配 0 个或多个路径（也就是用/分割的多级路径）。
- {spring:[a-z]+}：按照正则匹配[a-z]+，并且将其作为路径变量，变量名为"spring"。

例子如下：

- com/t?st.jsp：匹配 com/test.jsp，但也匹配 com/tast.jsp 或 com/txst.jsp。
- com/*.jsp：匹配 com 目录下所有.jsp 文件。
- com/**/test.jsp：匹配 com 目录以及子目录中所有.jsp 文件。
- org/springframework/**/*.jsp：匹配 org/springframework 目录及子目录下所有.jsp 文件。
- com/{filename:\\w+}.jsp：例如 com/目录下有 test.jsp，test.jsp 将会被匹配上，并且将 test 赋值给 filename 这一变量。

5.2.4 权限管理注解

以上示例示范了基于路径匹配的权限管理，一定程度上解决了权限管理的问题。但面对更细粒度的需求，基于路径匹配的方案显得捉襟见肘。这时候便需要使用上 Spring Security 的注解。

启用这一方案的方式同样很简单，在配置类上使用注解@EnableGlobalMethodSecurity 加上对应属性即可。

【示例 5-6】

示例如下：

```
@Configuration
@RequiredArgsConstructor
```

```
@EnableGlobalMethodSecurity(prePostEnabled = true, securedEnabled = true,
jsr250Enabled = true)
public class SecurityConfig extends WebSecurityConfigurerAdapter {

    //……

}
```

不同属性对应着不同种类注解的开关。开启之后可以在程序的任意一层使用以下注解，对方法进行更为具体的权限管理。

（1）@Secured，用以标注访问方法所需具备的权限。示例代码：

```
@Secured({"ROLE_ADMINISTRATOR", "ROLE_STAFF"})
public String secured() {
    //超级管理员与普通管理员都可以调用
    return "administrator,staff";
}
```

（2）@PreAuthority 和@PostAuthority，相对灵活并且支持 Spring 表达式。示例代码：

```
@PreAuthorize("hasAuthority('READ_AUTHORITY') AND hasAuthority('WRITE_AUTHORITY')")
public String preAuthorize() {
    //在方法执行前进行权限验证
    return "hey u.";
}

@PostAuthorize("hasRole('ADMINISTRATOR') OR hasRole('STAFF')")
public String postAuthorize() {
    //在方法执行后进行权限验证
    return "Oh,you got it.";
}
```

（3）@RolesAllowed、@PermitAll 和@DenyAll。示例代码：

```
@DenyAll
public String deny() {
    //所有访客都不可调用
    return "No entry.";
}

@PermitAll
public String permitAll() {
    //所有访客都可以调用
    return "Be my guest.";
}
@RolesAllowed({"ADMINISTRATOR", "STAFF"})
public String rolesAllowed() {
    //超级管理员与普通管理员都可以调用
    return "Top Secret.";
}
```

5.3 Session-Cookie

Web 应用开发大多是基于 HTTP 协议的。客户端与服务端在使用 HTTP 协议进行交互时，协议本身并不会记录客户端的信息或是状态。这意味着 HTTP 协议是一个"无状态"协议，而 Web 应用中的很多场景是需要保存用户状态的。网站的安全性解决方案就需要对状态进行验证，访问一个受保护的资源的第一步便是检查用户是否通过登录验证。这样的情况催生了许多用于保存用户状态的解决方案，Session-Cookie 就是其中最为常见的一种。

5.3.1 什么是 Session-Cookie

Session-Cookie 由 Session 与 Cookie 两部分组成，分别存储于服务端和客户端。Session 是一种用于记录用户的状态信息，被存储于服务端的数据。例如当前用户是否登录，身份与权限是怎样的，这类数据大多会存储于 Session 当中。

Cookie（或者被称作 HTTP Cookie）则是一小片被存储在浏览器的数据，往往在浏览一个网站时便会产生。本身作用在于可以在访问网站的时候，记录一些状态信息到客户端。可以起到提升系统性能并且提高用户体验。在 Session-Cookie 体系中，Cookie 主要用于记录一串与 Session 关联的标识。该标识往往被称作"session-id"。

Session 与 Cookie 不同的是，Cookie 是一个真实存在的具体实现，而 Session 是一个相对抽象的概念，不同的语言与开发框架对 Session 有不同的实现方式。这两者共同作用，使得用户即便是在使用"无状态"协议与服务端交互，依旧能在不同页面共享一些信息。这个交互流程如图 5.4 所示。

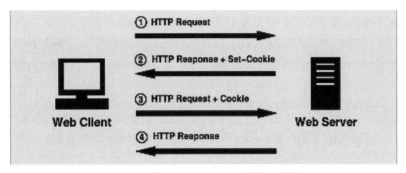

图 5.4 基于 Session-Cookie 的交互流程

- 客户发送请求到服务端。
- 服务端收到请求之后便会在内部创建 Session，之后返回响应。在响应头中，包含 Set-Cookie 信息，其中内容就包含 Session-id。
- 当客户端收到包含 Set-Cookie 的响应之后，后续的所有请求就会带上 Cookie。
- 服务端可以根据 Cookie 中的 Session-id 让每个请求与 Session 逐一对应，实现请求间的状态共享。

5.3.2 使用 Spring Session 管理 Session

在 Spring Boot 开发中，Session-Cookie 默认由 Web 容器（例如 Tomcat）维护。例如，访问一个 Web 容器是 Tomcat 的 Spring Boot 应用，默认返回的响应头中会设置一条键为 JSESSIONID 的 Cookie，如图 5.5 所示。

```
HTTP/1.1 200
X-Content-Type-Options: nosniff
X-XSS-Protection: 1; mode=block
Cache-Control: no-cache, no-store, max-age=0, must-revalidate
Pragma: no-cache
Expires: 0
X-Frame-Options: DENY
Set-Cookie: JSESSIONID=7402A8057DD241D1D35B0FFB4B505D3F; Path=/; HttpOnly
Content-Type: application/json;charset=utf-8
Transfer-Encoding: chunked
Date: Wed, 23 Sep 2020 15:17:37 GMT
Keep-Alive: timeout=60
Connection: keep-alive
```

图 5.5　Tomcat 包含 Set-Cookie 的响应

Session 默认情况下是保存于服务端的 JVM 内存当中。在该条件下当服务端程序重启时，Session 将会丢失。这也意味着用户会集体掉线，如果要继续操作则需要重新登录。在某些场景下，该情况会严重影响用户的使用体验。这不得不将 Session 的管理从 JVM 内存中剥离出来。

Spring Session 便提供了一套方案用于解决 Session 管理的问题，使其不依赖于特定的应用程序容器实现 Session 集群化。下面将讲解 Spring Boot 整合 Spring Session 的步骤。

（1）更新依赖

整合 Spring Session 的第一个步骤是更新依赖。Spring Session 可选的容器类型有四种，分别是 Redis、MongoDB、JDBC 和 HAZELCAST。选择好对应的容器类型后，再根据类型选择对应的依赖项。比如选择 Redis 作为 Session 的外部容器，则需要引入 spring-session-data-redis。

pom.xml：

```xml
<dependency>
    <groupId>org.springframework.boot</groupId>
    <artifactId>spring-boot-starter-data-redis</artifactId>
</dependency>
```

（2）Spring Boot 配置

在添加完依赖之后，可以着手开始进行针对 Spring Boot 的配置。这一步所要做的也很简单，依然以 Redis 作为 Session 的外部容器为例，在 application.yml 中加上如下内容：

```yaml
spring:
  session:
    store-type: redis
```

得益于 Spring Boot 的自动配置，加上该配置项之后等同于使用了注解

@EnableRedisHttpSession。这将创建名为 springSessionRepositoryFilter 的过滤器。该过滤器负责将容器中的 HttpSession 替换为 Spring Session。

另外还有一些高级配置：

```
spring:
  session:
    #会话超时。如果未指定持续时间后缀，则使用秒
    timeout: 18000
    redis:
      #ON_SAVE 和 IMMEDIATE
      flush-mode: ON_SAVE
      #用于存储 Session 的命名空间
      namespace: spring:session
```

（3）配置 Redis 连接

```
redis:
  #redis 域名
  hostname: localhost
  #redis 端口
  port: 6379
  #redis 密码
  password: NlzWZLvvCF5Gzzby
```

（4）检验结果

启动程序，并且成功登录之后，可以通过 Redis 的 GUI 管理工具查看对应命名空间下的存储情况，如图 5.6、图 5.7 所示。

图 5.6　Spring Session 存储于 Redis 的内容列表

ID (Total: 5)	Key	Value
1	maxInactiveInterval	\xac\xed\x00\x05sr\x00\x11java.lang.Integer\x12\xe2\xa0\xa4\xf7\x81\x878\x02\x00\x01I\x00\...
2	lastAccessedTime	\xac\xed\x00\x05sr\x00\x0ejava.lang.Long;\x8b\xe4\x90\xcc\x8f#\xdf\x02\x00\x01J\x00\x05val\...
3	sessionAttr:SPRING_SECURITY_SAVED_REQUEST	\xac\xed\x00\x05sr\x00Aorg.springframework.security.web.savedrequest.DefaultSavedRequest\...
4	sessionAttr:SPRING_SECURITY_CONTEXT	\xac\xed\x00\x05sr\x00=org.springframework.security.core.context.SecurityContextImpl\x00\x\...
5	creationTime	\xac\xed\x00\x05sr\x00\x0ejava.lang.Long;\x8b\xe4\x90\xcc\x8f#\xdf\x02\x00\x01J\x00\x05val\...

图 5.7　Session 内容

尝试将其删除，可以观察到客户端会处于登出状态。

5.3.3 Session 并发配置

Spring Security 中默认对 session 的并发数并没有限制。换句话说，一个用户的账号和密码在默认条件下可以供任意数量的客户端登录。这种情况在某些场景下是难以接受的，例如，一些付费的视频，对于同账户的客户端在线数量会进行限制。Spring Security 对于这样的场景也提供了支持。

【示例 5-7】

首先定义一个 SessionInformationExpiredStrategy 的实现作为失效策略，用于返回异常信息。ParallelismSessionExpiredStrategy.java：

```
@RequiredArgsConstructor
public class ParallelismSessionExpiredStrategy implements
SessionInformationExpiredStrategy {

    private final ObjectMapper objectMapper;

    @Override
    public void onExpiredSessionDetected(SessionInformationExpiredEvent event)
throws IOException, ServletException {
        //返回异常信息
event.getResponse().setContentType("application/json;charset=UTF-8");
event.getResponse().getWriter().write(objectMapper.writeValueAsString(R.failed
("已达到并发上限")));
    }
}
```

SecurityConfig 中加上 sessionManagement 配置，SecurityConfig.java：

```
@Configuration
@RequiredArgsConstructor
@EnableGlobalMethodSecurity(prePostEnabled = true, securedEnabled = true,
jsr250Enabled = true)
public class SecurityConfig extends WebSecurityConfigurerAdapter {

    @Override
    protected void configure(HttpSecurity http) throws Exception {
        http
            //……
            .sessionManagement()
            //并发 Session 上限为 1
            .maximumSessions(1)
            //达到上限后是否阻止登录
            .maxSessionsPreventsLogin(true)
            //失效 Session 策略
            .expiredSessionStrategy(new
ParallelismSessionExpiredStrategy(mapper));
    }
            //……
```

}

其中，maxSessionsPreventsLogin 用于控制达到 Session 并发上限后的策略，True 代表将阻止后续的登录操作，False 代表将会使之前登录的 Session 强制失效。另外，失效 Session 的请求结果将按照 expiredSessionStrategy 中配置的策略进行返回。

5.3.4 强制下线

Session-Cookie 的特点在于服务端可以监听并控制用户的会话状态。这为会话管理模块的开发提供了很大便利。会话管理中强制下线会是一个常见需求，如果一个 Web 应用中出现用户在进行违规操作，可以使用强制下线功能为系统提供保护。

【示例 5-8】

为了实现强制下线功能，需要在 SecurityConfig 中进行进一步配置。SecurityConfig.java：

```java
@Configuration
@RequiredArgsConstructor
@EnableGlobalMethodSecurity(prePostEnabled = true, securedEnabled = true, jsr250Enabled = true)
public class SecurityConfig extends WebSecurityConfigurerAdapter {

    @Override
    protected void configure(HttpSecurity http) throws Exception {
        http
            //……
            .sessionManagement()
            //并发 Session 上限为 1
            .maximumSessions(1)
            //达到上限后是否阻止登录
            .maxSessionsPreventsLogin(true)
            //失效 Session 策略
            .expiredSessionStrategy(new ParallelismSessionExpiredStrategy(mapper))
            .sessionRegistry(sessionRegistry());
    }

    @Bean
    public SessionRegistry sessionRegistry() {
        //用于访问 Session
        return new SessionRegistryImpl();
    }

    @Bean
    public static ServletListenerRegistrationBean httpSessionEventPublisher() {
        //用于告诉 Spring 将 Session 信息存储于 sessionRegistry
        return new ServletListenerRegistrationBean(new HttpSessionEventPublisher());
    }
    //……
}
```

配置完成后便可以着手实现对应的工具类，SessionUtils.java：

```java
@Component
@RequiredArgsConstructor
public class SessionUtils {

    private final SessionRegistry sessionRegistry;

    public void expireUserSessions(String username) {
        for (Object principal : sessionRegistry.getAllPrincipals()) {
            if (principal instanceof SecurityProperties.User) {
                UserDetails userDetails = (UserDetails) principal;
                //遍历 sessionRegistry 中的 principal 找到对应用户的 Session
                if (userDetails.getUsername().equals(username)) {
                    for (SessionInformation information : sessionRegistry.getAllSessions(userDetails, true)) {
                        //让 session 立刻失效
                        information.expireNow();
                    }
                }
            }
        }
    }
}
```

当需要使某个用户强制下线时，调用 SessionUtils 的 expireUserSessions() 命令即可。

5.4　JWT（JSON Web Token）

在 5.3 节介绍了基于 Session-Cookie 模式的身份验证，这类模式的特点在于后端会维护用户的状态。在常见身份验证方式中，有一种方式与 Session-Cookie 有着鲜明差异，它不用于后端维护用户状态，这一方式正是本节将介绍的对象——JWT。

5.4.1　关于 JWT

JWT 是一个开放标准（RFC 7519），它定义了一种紧凑且自包含的方式，用于在各端之间将安全信息以 JSON 对象的形式进行传递。由于此信息是经过数字签名的，因此可以被验证和信任。另外，还可以使用 HMAC 或者 RSA 算法对 JWT 进行加密。

JWT 以紧凑的三部分组成，各部分以（.）分隔，形如 xxxx.yyy.zzz。这三部分分别是：

- Header（标头）。通常由两部分组成，令牌的类型（即 JWT）和所使用的签名算法，例如 HMAC SHA256 或 RSA。
- Payload（负载）。负载中包含不同类型的声明（Claims）。有已注册声明（预定义声明，包含签发人、超时时间等信息）、公有声明和私有声明（由签发者用于在各端传递安全信息的声明，可随意定义）。
- Signature（签名）。基于标头中定义的加密算法进行加密的签名。例如，加密算法为

HMACSHA256 的话，签名过程将会是：

```
HMACSHA256(
  base64UrlEncode(header) + "." +
  base64UrlEncode(payload), secret)
```

签名的作用在于保证消息在传输过程中是未被更改的，并且如果使用私钥进行签名，还可以对发送者身份进行进一步的认证。

基于 JWT 加密并且可携带安全信息的特点，后端可以将原本保存于 Session 中的信息保存至 JWT 的私有声明，然后要求客户端在每次请求中都携带 JWT。以这种方式实现后端服务的无状态化。

JWT 带来后端无状态化的同时，也带来了不少好处：

（1）跨域与 CORS

传统的认证方式中，Cookies 是默认只能用于单个域名和子域名间进行消息传递。如果碰到域名不同的情况（也就是跨域），整个处理过程会变得繁琐。JWT 不依赖 Cookie，大多以 Authorization: Bearer {JWT}这样的格式作为请求头进行传递，不会被跨域请求所影响。

（2）跨平台解决方案

同样是受限于 Cookies 的特性，移动平台与 Cookies 并不能很好地融合，存在诸多限制。JWT 相比 Cookie-Session 更适合作为跨平台的认证解决方案。

（3）方便扩展，提升后端性能

Session-Cookie 的验证过程中基本都需要对存储模块进行访问，特别在于分布式架构中的认证过程，后端服务需要对外部存储进行访问，其中必然存在数据传输带来的时间损耗。JWT 的验证过程中仅需要根据加密算法对 JWT 进行验签解密。相对来说，不会有扩展和性能方面的困扰。

5.4.2 JWT 工作流程

通常情况下 JWT 通过访问一个用于登录的 URL 来获取。登录成功后，客户端将会在之后的请求的每个请求体中，以 Authorization: Bearer {JWT}的格式将 JWT 设置为请求头。服务端将会对 JWT 内容进行验签解密，并做出相应的响应。JWT 的工作流程图如图 5.8 所示。

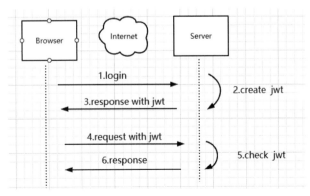

图 5.8 JWT 工作流程示意图

5.4.3 Spring Security 集成 JWT

基于 JWT 的认证模式需要开发人员对 Spring Security 默认的认证模式进行一定程度的调整。

【示例 5-9】

（1）引入依赖

基于 JWT 的认证需要用到 Spring Security 之外的依赖，如下：

```xml
<dependency>
    <groupId>io.jsonwebtoken</groupId>
    <artifactId>jjwt</artifactId>
    <version>0.9.1</version>
</dependency>
```

该依赖提供了 JWT 认证所需的验证、读写声明等基本功能。

（2）创建基础工具类

创建一个 JwtTokenUtil，其中封装了若干 JWT 相关的方法供后续功能开发使用。
JwtTokenUtil.java：

```java
@Component
@RequiredArgsConstructor
public class JwtTokenUtil implements Serializable {

    private static final long serialVersionUID = 1L;
    /**
     * token 有效期，单位为秒
     */
    private static final long JWT_TOKEN_VALIDITY = 5 * 60 * 60;
    @Value("${jwt.secret}")
    private String secret;
    private final ObjectMapper objectMapper;

    public String getUsernameFromToken(String token) {
        //从 JWT 中获取用户名
        return getClaimFromToken(token, Claims::getSubject);
    }

    public Date getExpirationDateFromToken(String token) {
        //从 JWT 中获取过期时间
        return getClaimFromToken(token, Claims::getExpiration);
    }

    public <T> T getClaimFromToken(String token, Function<Claims, T> claimsResolver) {
        final Claims claims = getAllClaimsFromToken(token);
        return claimsResolver.apply(claims);
    }

    private Claims getAllClaimsFromToken(String token) {
        //通过秘钥获取 JWT 中所有 Claims 信息
        return
```

```
Jwts.parser().setSigningKey(secret).parseClaimsJws(token).getBody();
    }

    private Boolean isTokenExpired(String token) {
        //确认Token是否过期
        final Date expiration = getExpirationDateFromToken(token);
        return expiration.before(new Date());
    }

    public String generateToken(UserDetails userDetails) throws
JsonProcessingException {
        //为某用户生成Token
        Map<String, Object> claims = new HashMap<>();
        //……此处可以根据需求在Claims中新增用户状态信息,此处将Principle以
JSONString的格式加入到Claims中属于可选项
        JwtUser principal = new JwtUser()
                .setUsername(userDetails.getUsername())
                .setPassword(userDetails.getPassword())
                .setAuthorities(userDetails.getAuthorities()
                    .stream()
                    .map(GrantedAuthority::getAuthority)
                    .collect(Collectors.toList()));
        claims.put(Constants.PRINCIPAL,
objectMapper.writeValueAsString(principal));
        return doGenerateToken(claims, userDetails.getUsername());
    }

    private String doGenerateToken(Map<String, Object> claims, String subject)
{
        //生成Token的大致步骤:
        //1.定义令牌的声明信息,例如主体、过期时间
        //2.使用HS512算法与配置好的秘钥对JWT进行加密
        //3.对JTW进行压缩
        return Jwts
                .builder()
                .setClaims(claims)
                .setSubject(subject)
                .setIssuedAt(new Date(System.currentTimeMillis()))
                .setExpiration(new Date(System.currentTimeMillis() +
JWT_TOKEN_VALIDITY * 1000))
                .signWith(SignatureAlgorithm.HS512, secret)
                .compact();
    }

    public Boolean validateToken(String token, String username) {
        //验证Token
        return Objects.equals(getUsernameFromToken(token), username)
                && !isTokenExpired(token);
    }
}
```

配置文件application.yml中需要配置加密所需的秘钥,如下所示:

```
jwt:
  secret: 2g3q6f&jZmvNV6KT!&5E11udkRbbH!LZ
```

（3）创建登录控制器

登录控制器将验证账号和密码的正确性，并且签发对应的 JWT。示例代码 JwtAuthenticationController.java：

```java
@RestController
@RequiredArgsConstructor
public class JwtAuthenticationController {

    private final AuthenticationManager authenticationManager;
    private final JwtTokenUtil jwtTokenUtil;
    private final UserDetailsService userDetailsService;


    @PostMapping("/authenticate")
    public R<?> createAuthenticationToken(@RequestBody UserDto authenticationRequest, HttpServletResponse response) throws Exception {
        authenticate(authenticationRequest.getUsername(), authenticationRequest.getPassword());
        UserDetails userDetails = userDetailsService.loadUserByUsername(authenticationRequest.getUsername());
        String token = jwtTokenUtil.generateToken(userDetails);
        setCookie(token, response);
        return R.ok(token);
    }

    private void authenticate(String username, String password) throws Exception {
        //并不在该类进行验证用户名密码，而是委托给 AuthenticationManager
        try {
            authenticationManager.authenticate(new UsernamePasswordAuthenticationToken(username, password));
        } catch (DisabledException e) {
            throw new Exception("USER_DISABLED", e);
        } catch (BadCredentialsException e) {
            throw new Exception("INVALID_CREDENTIALS", e);
        }
    }

    private void setCookie(String token, HttpServletResponse response) {
        //将 JWT 写入 Cookie，用于减少客户端对 Header 的操作(该步骤为可选项)
        Cookie cookie = new Cookie(Constants.JWT, token);
        response.addCookie(cookie);
    }
}
```

（4）Spring Security 配置

较之 Session Cookie，JWT 的配置会存在一定差异。需要全局配置 AuthenticationManager 的 userDetailsService，还要将 Session 创建方式修改为无状态模式。示例代码 SecurityConfig.java：

```java
@Configuration
@RequiredArgsConstructor
@EnableGlobalMethodSecurity(prePostEnabled = true, securedEnabled = true, jsr250Enabled = true)
```

```java
public class SecurityConfig extends WebSecurityConfigurerAdapter {
    private final ObjectMapper mapper;
    private final DataSource dataSource;
    private final UserDetailsService userDetailsService;
    private final JwtAuthenticationEntryPoint jwtAuthenticationEntryPoint;
    private final JwtRequestFilter jwtRequestFilter;

    @Autowired
    public void configureGlobal(AuthenticationManagerBuilder auth) throws Exception {
        //为 AuthenticationManager 配置 userDetailService 与 passwordEncoder
        auth.userDetailsService(userDetailsService).passwordEncoder(bCryptPasswordEncoder());
    }

    @Bean
    public PasswordEncoder bCryptPasswordEncoder() {
        return new BCryptPasswordEncoder();
    }


    @Bean
    @Override
    public AuthenticationManager authenticationManagerBean() throws Exception {
        return super.authenticationManagerBean();
    }

    @Override
    protected void configure(HttpSecurity http) throws Exception {
        http
                //开启登录配置
                .authorizeRequests()
                //表示 account 接口的访问不受限制
                .antMatchers("/account").permitAll()
                .antMatchers("/authenticate").permitAll()
                //表示所有接口,登录之后就能访问
                .anyRequest().authenticated()
                .and()
                .logout()
                .logoutUrl("/logout")
                //登出成功的处理程序
                .logoutSuccessHandler((req, resp, authentication) -> {
                    resp.setContentType("application/json;charset=utf-8");
                    PrintWriter out = resp.getWriter();
                    out.write(mapper.writeValueAsString(R.<Void>ok(null)));
                    out.flush();
                })
                .permitAll()
                .and()
                .httpBasic()
                .disable()
```

```
                        .csrf()
                        .disable()
                        .exceptionHandling().authenticationEntryPoint(jwtAuthenticati
onEntryPoint)
                        .and()
                        .sessionManagement()
                        .sessionCreationPolicy(SessionCreationPolicy.STATELESS);

            //将jwtRequestFilter插入到过滤器链的头部用来验证每个请求的Token
            http.addFilterBefore(jwtRequestFilter,
UsernamePasswordAuthenticationFilter.class);
        }

        @Bean
        public PersistentTokenRepository persistentTokenRepository() {
            JdbcTokenRepositoryImpl tokenRepository = new
JdbcTokenRepositoryImpl();
            tokenRepository.setDataSource(dataSource);
            return tokenRepository;
        }
    }
```

（5）认证异常处理

JwtAuthenticationEntryPoint.java：

```
    @Component
    public class JwtAuthenticationEntryPoint implements AuthenticationEntryPoint
{
        @Override
        public void commence(HttpServletRequest request, HttpServletResponse
response, AuthenticationException authException) throws IOException,
ServletException {
            response.sendError(HttpServletResponse.SC_UNAUTHORIZED,
"Unauthorized");
        }
    }
```

（6）创建用于验证 JWT 的过滤器

JWT 的验证依赖于根据需求自定义的 OncePerRequestFilter 实现，示例代码 JwtRequestFilter.java：

```
    @Component
    @RequiredArgsConstructor
    public class JwtRequestFilter extends OncePerRequestFilter {

        private final JwtTokenUtil jwtTokenUtil;
        private final ObjectMapper objectMapper;

        @Override
        protected void doFilterInternal(HttpServletRequest request,
HttpServletResponse response, FilterChain filterChain) throws ServletException,
IOException {
            String requestTokenHeader = request.getHeader("Authorization");
```

```java
            String username = null;
            String jwtToken = null;

            // JWT 一般是"Bearer token"这样的格式，去除 Bearer 前缀，只留下 token 本身
            String bearerPrefix = "Bearer ";
            if (requestTokenHeader != null &&
requestTokenHeader.startsWith(bearerPrefix)) {
                jwtToken = requestTokenHeader.substring(7);
            } else {
                logger.warn("JWT Token does not begin with Bearer String");
                if (request.getCookies() != null) {
                    for (Cookie cookie : request.getCookies()) {
                        if (Constants.JWT.equals(cookie.getName())) {
                            jwtToken = cookie.getValue();
                        }
                    }
                }
            }
            try {
                username = jwtTokenUtil.getUsernameFromToken(jwtToken);
            } catch (IllegalArgumentException e) {
                System.out.println("Unable to get JWT Token");
            } catch (ExpiredJwtException e) {
                System.out.println("JWT Token has expired");
            }


            // 验证 token
            boolean setAuthentication = username != null &&
SecurityContextHolder.getContext().getAuthentication() == null &&
jwtTokenUtil.validateToken(jwtToken, username);
            if (setAuthentication) {
                //如果 token 有效，则进行手动授权并将其设置到 Spring Security 的上下文中
                String userJson = jwtTokenUtil.getClaimFromToken(jwtToken, (c) ->
    c.get(Constants.PRINCIPAL, String.class));
                JwtUser jwtUser = objectMapper.readValue(userJson, JwtUser.class);
                User userDetails = new User(jwtUser.getUsername(),
jwtUser.getPassword(), jwtUser.getAuthorities()
                        .stream()
                        .map(SimpleGrantedAuthority::new)
                        .collect(Collectors.toList()));
                UsernamePasswordAuthenticationToken
usernamePasswordAuthenticationToken = new UsernamePasswordAuthenticationToken(
userDetails, null, userDetails.getAuthorities());
                usernamePasswordAuthenticationToken.setDetails(new
WebAuthenticationDetailsSource().buildDetails(request));

                SecurityContextHolder.getContext().setAuthentication(usernamePass
wordAuthenticationToken);
            }
            filterChain.doFilter(request, response);
        }
    }
```

(7)安全信息

私有声明中可以存储用于安全认证的安全信息，本示例中安全信息将存储用户的用户名、加密后的密码以及权限，JwtUser.java 示例代码：

```java
@Accessors(chain = true)
@Setter
@Getter
public class JwtUser implements Serializable {

    private static final long serialVersionUID = -5954106717635750452L;
    private String username;
    private String password;
    private List<String> authorities;

}
```

5.5　OAuth 2.0

第三方应用授权在互联网服务中被广泛地运用。在这一功能的背后，大多会使用到 OAuth 这一授权技术。OAuth 是一个用于授权的网络标准，目前主流的版本为 2.0 版。本节将介绍 OAuth 2.0 的基本概念以及 Spring Boot 对该技术的集成。

5.5.1　OAuth 2.0 简介

使用 OAuth 2.0 这一授权标准，可以让第三方应用程序获取用户在某一网络服务（例如 QQ、GitHub）上有限的账户访问权限。比如，使用 QQ 的第三方授权可以获取用户的 QQ 昵称头像等信息。实现方式是通过将身份验证这一环委托给承载用户账户的服务器（通常被称作认证服务器，Authorization Server），并授权第三方应用（通常被称作资源服务器，Resource Server）访问用户账户。

OAuth 2.0 中涉及四个角色之间的信息交换，分别是：

- 用户：用户拥有授权/资源服务器上的账户，在流程中授予客户端访问账户的权限。
- 授权服务器：负责验证用户身份，向第三方应用颁发令牌。
- 资源服务器：负责保管用户账户。
- 第三方应用（客户端）：想要访问用户账户的应用程序。

OAuth 2.0 工作流程如图 5.9 所示。

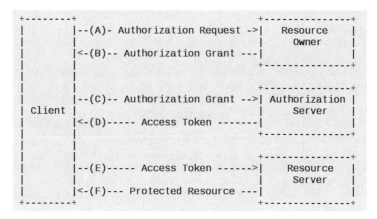

图 5.9 OAuth 2.0 工作流程

上图摘自 RFC 6749，对其进一步说明：

（A）用户打开客户端后，客户端要求用户给予授权。
（B）用户同意给予客户端授权。
（C）用户基于上一步的授权向认证服务器申请令牌。
（D）认证服务器对客户端成功认证后发放令牌。
（E）客户端使用令牌，向资源服务器申请获取资源。
（F）资源服务器确认令牌无误，同意向客户端开放资源。

5.5.2 授权模式

OAuth 2.0 提供四种不同的授权模式以适应各种应用场景：

- 授权码模式
- 密码模式
- 简化模式
- 客户端模式

1. 授权码模式

授权码模式是最常用的授权模式，是针对服务端应用程序优化过的方案。在该模式下，源码不会公开，客户端秘钥可以保证机密性。这是一个基于重定向的流程，意味着应用程序必须能与用户代理（即浏览器）进行交互，并且能够接收通过用户代理路由的 API 授权代码。流程如图 5.10 所示。

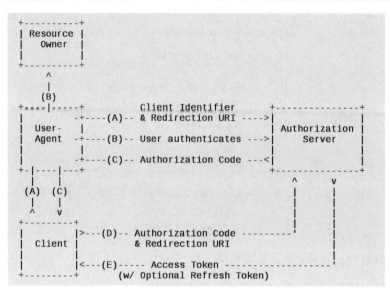

图 5.10　授权码模式

流程说明如下：

（A）用户访问客户端，后者将前者导向认证服务器。

（B）用户选择是否给予客户端授权。

（C）假设用户给予授权，认证服务器将用户导向客户端事先指定的"重定向 URI"，同时附上授权码。

（D）客户端收到授权码，附上"重定向 URI"，向认证服务器申请令牌。这一步在服务器间完成交互过程，对用户不可见。

（E）认证服务器核对授权码和重定向 URI，之后向客户端发送访问令牌和更新令牌。

2. 密码模式

使用密码模式授权，用户可以直接向应用程序提供其凭证（用户名和密码），该应用程序使用凭证从服务器获取访问令牌。仅当其他流程不可行时，才考虑在授权服务器上启用此授予类型。同样，仅应在用户信任该应用程序（例如系统程序）时使用，流程如图 5.11 所示。

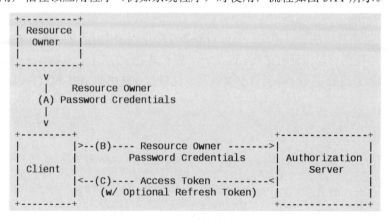

图 5.11　密码模式

流程说明如下：

（A）用户向客户端提供用户名和密码。
（B）客户端将用户名和密码发给认证服务器，向后者请求令牌。
（C）认证服务器确认无误后，向客户端提供访问令牌。

3. 简化模式

该模式多用于移动互联网或网站，其中客户端机密性无法保证。简化模式与授权码模式相似，也是基于重定向，但访问令牌会由用户代理转发给应用程序，因此令牌处于公开可见的状态。并且流程中不对应用程序的身份进行验证，仅依赖重定向 URI 需要事先注册来实现验证的目的，流程如图 5.12 所示。

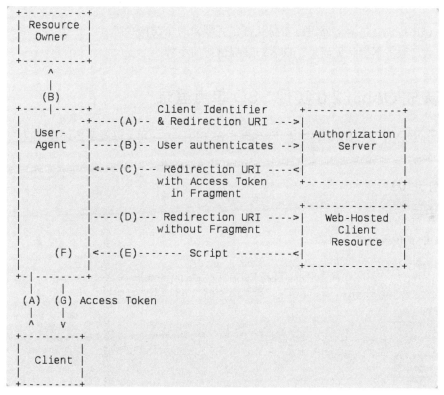

图 5.12　简化模式

流程说明如下：

（A）客户端将用户导向认证服务器。
（B）用户决定是否给予客户端授权。
（C）假设用户给予授权，认证服务器将用户导向客户端指定的"重定向 URI"，并在 URI 的 Hash 部分包含了访问令牌。
（D）浏览器向资源服务器发出请求，其中不包括上一步收到的 Hash（哈希）值。
（E）资源服务器返回一个网页，其中包含的代码可以获取 Hash 值中的令牌。
（F）浏览器执行上一步获得的脚本，提取出令牌。

（G）浏览器将令牌发给客户端。

4. 客户端模式

该模式下客户端将代替用户以自己的名义向服务器提交认证，流程如图 5.13 所示。

图 5.13　客户端模式

流程说明如下：

（A）客户端向认证服务器进行身份认证，并要求一个访问令牌。

（B）认证服务器确认无误后，向客户端提供访问令牌。

5.5.3　集成 OAuth 2.0 实现 SSO 单点登录

该集成示例基于授权码模式实现，分为客户端（第三方应用）以及服务端（授权/资源服务器）两大部分。

【示例 5-10】

1. 客户端

客户端的实现需要如下依赖：

```xml
<dependency>
    <groupId>org.springframework.boot</groupId>
    <artifactId>spring-boot-starter-web</artifactId>
</dependency>
<dependency>
    <groupId>org.springframework.boot</groupId>
    <artifactId>spring-boot-starter-security</artifactId>
</dependency>
<dependency>
    <groupId>org.springframework.security.oauth.boot</groupId>
    <artifactId>spring-security-oauth2-autoconfigure</artifactId>
    <version>2.0.1.RELEASE</version>
</dependency>
<dependency>
    <groupId>org.springframework.boot</groupId>
    <artifactId>spring-boot-starter-thymeleaf</artifactId>
</dependency>
<dependency>
    <groupId>org.thymeleaf.extras</groupId>
    <artifactId>thymeleaf-extras-springsecurity4</artifactId>
</dependency>
```

客户端不需要编写过多业务代码，示例中的重点在于应用的安全性配置，

UiSecur-ityConfig.java:

```
@Configuration
@EnableOAuth2Sso
public class UiSecurityConfig extends WebSecurityConfigurerAdapter {

    @Override
    public void configure(HttpSecurity http) throws Exception {
        http.antMatcher("/**")
            .authorizeRequests()
            .antMatchers("/", "/login**")
            .permitAll()
            .anyRequest()
            .authenticated();
    }
}
```

配置中的@EnableOAuth2Sso 注解用于开启单点登录功能。配置的内容用于分配请求的处理范围。

```
application.yml:
server:
   port: 8082
   servlet:
      context-path: /ui
security:
  basic:
    enabled: false
  oauth2:
    client:
      clientId: 42660590
      clientSecret: zG0k58iYEEyb60eP
      #该配置项用于获取访问令牌的地址
      accessTokenUri: http://localhost:8081/auth/oauth/token
      #该配置项用于将用户重定向至授权地址
      userAuthorizationUri: http://localhost:8081/auth/oauth/authorize
resource:
   #该配置项用于定义获取用户信息的地址
      userInfoUri: http://localhost:8081/auth/user/current
spring:
  thymeleaf:
    cache: false
```

另外还有两个前端模板用于展示，文件名为 index.html 与 securePage.html。

index.html：

```
<h1>Spring Security SSO</h1>
<a href="securedPage">Login</a>
```

securePage.html：

```
<h1>Secured Page</h1>
Welcome,
<span th:text="${#authentication.name}">Name</span>
```

2. 授权/资源服务

本示例中授权服务于资源服务实际上为一个可部署单元,依赖项如下:

```xml
<dependency>
    <groupId>org.springframework.boot</groupId>
    <artifactId>spring-boot-starter-web</artifactId>
</dependency>
<dependency>
    <groupId>org.springframework.security.oauth</groupId>
    <artifactId>spring-security-oauth2</artifactId>
    <version>2.3.3.RELEASE</version>
</dependency>
```

入口类 AuthorizationServerApplication.java:

```java
@SpringBootApplication
@EnableAuthorizationServer
@EnableResourceServer
public class AuthorizationServerApplication extends SpringBootServletInitializer {
    public static void main(String[] args) {
        SpringApplication.run(AuthorizationServerApplication.class, args);
    }
}
```

其中@EnableAuthorizationServer 与@EnableResourceServer 两个注解分别用来开启授权服务与资源服务。

授权服务配置类 AuthServerConfig.java:

```java
@Configuration
public class AuthServerConfig extends AuthorizationServerConfigurerAdapter {
    @Autowired
    private BCryptPasswordEncoder passwordEncoder;
    @Override
    public void configure(
      AuthorizationServerSecurityConfigurer oauthServer) throws Exception {
        oauthServer.tokenKeyAccess("permitAll()")
          .checkTokenAccess("isAuthenticated()");
    }

    @Override
    public void configure(ClientDetailsServiceConfiqurer clients) throws Exception {
        //使用 inMemory 方式注册客户端,另外还可以使用 JDBC 方式进行注册
        clients.inMemory()
            .withClient("42660590")
            .secret(passwordEncoder.encode("zG0k58iYEEyb60eP "))
            //授权模式为授权码模式
            .authorizedGrantTypes("authorization_code")
            .scopes("user_info")
            .autoApprove(true)
            //注册重定向 URI
            .redirectUris("http://localhost:8082/ui/login","http://localhost:8083/ui2/login");
```

```
    }
}
```

application.yml：

```
server:
  port: 8081
  servlet:
context-path: /auth
```

安全性配置 SecurityConfig.java：

```
@Configuration
@Order(1)
public class SecurityConfig extends WebSecurityConfigurerAdapter {

    @Override
    protected void configure(HttpSecurity http) throws Exception {
        http.requestMatchers()
            .antMatchers("/login", "/oauth/authorize")
            .and()
            .authorizeRequests()
            .anyRequest().authenticated()
            .and()
            .formLogin().permitAll();
    }

    @Override
    protected void configure(AuthenticationManagerBuilder auth) throws Exception {
        auth.inMemoryAuthentication()
            .withUser("someone")
            .password(passwordEncoder().encode("1234567"))
            .roles("USER");
    }

    @Bean
    public BCryptPasswordEncoder passwordEncoder(){
        return new BCryptPasswordEncoder();
    }
}
```

最后声明用于返回用户信息的控制器 UserController.java：

```
@RestController
public class UserController {
    @GetMapping("/user/me")
    public Principal user(Principal principal) {
        return principal;
    }
}
```

第 6 章

自动化测试

自动化测试是一种软件测试技术，通过使用特殊的自动化测试工具执行测试用例。与之对应的是手工测试，是由坐在计算机前的开发人员或测试人员，通过手动输入的方式执行测试步骤以达到测试的目的。使用该技术的目标在于减少软件开发阶段手动运行测试用例的数量，从而提高测试效率与质量。当然，自动化测试并不能完全替代手工测试，合理地结合使用两者可以更好地保证代码质量。

本章将围绕自动化测试的主题，介绍该技术在 Spring Boot 程序开发中的运用，涉及的知识点如下：

- 基于 JUnit 的单元测试
- 使用 Hamcrest 为测试用例断言
- 基于 Spring Boot Test 进行分层级的测试
- 测试中的 Mock

6.1 单元测试

单元测试在自动化测试技术当中属于最为基础的测试环节。Spring Boot 程序开发过程中，通常依赖 JUnit 对程序进行单元测试。本节将介绍单元测试的一些概念以及如何使用 JUnit 完成单元测试。

6.1.1 测试金字塔

如果想要更好地完成所有自动化测试，那么有一个概念需要了解一下——测试金字塔模型

（Test Pyramid），如图 6.1 所示。

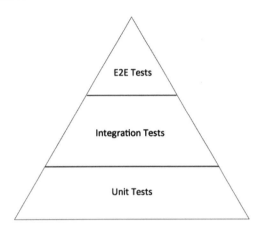

图 6.1　测试金字塔

该模型将自动化测试按粒度划分为不同类型，从下至上依次为单元测试（Unit Test）、集成测试（Integration Test）与端对端测试（End To End Test）。这一概念最初由迈克·科恩（Mike Cohn）提出，是一种非常直观易懂的视觉隐喻。它告诉开发人员，需要根据层级划分测试内容，并且不同层级中的测试数量大概需要如何分配。在这三个层级中，下方的测试层相对更独立，执行效率相对高；越往上集成的程度就越高，执行效率相对低。

虽然图中明确划分了三种测试层级，但现实的开发过程中会遇到的情况千变万化，所需要编写的测试层级会相应变多或者变少。一切需要视情况而定，切忌生搬硬套。比如图 6.1 中的三种测试名称，在不同开发团队中也会在确定好粒度与边界后使用不同名称用以代替。一般将仅针对单个方法或者单个类的测试称作小型测试，将针对服务或者模块的测试称作中型测试……因为缺少明确与严格的规定，最好的选择是记住其中想要传递的两点信息：

- 编写不同粒度的测试。
- 粒度越大的测试，所应编写的用例应该越少。

本节将介绍的正是处于测试金字塔最底层的单元测试，通常指的是针对最小可测试单元的测试。换句话来说就是所测试单元与外界的依赖尽可能为零。符合该条件的测试场景，可以通过下文将要介绍的手段编写测试用例。

6.1.2　JUnit 基础

JUnit 是 Java 编程中的一个测试框架，用于编写可重复运行的自动化测试。特点如下：

- 简洁优雅，方便上手。
- 通过注解来识别测试方法。
- 通过断言的方式测试预期结果。
- 可以自动运行、检查结果并且提供即时反馈，减少了人工整理测试结果这一过程。

注解在 JUnit 中扮演着重要角色，JUnit 的使用基本围绕着注解展开。注解如表 6.1 所示。

表 6.1 JUnit 中的注解

注解	描述
@Test	作用于方法，标记一个方法即可作为一个测试用例
@Before	作用于方法，该方法必须在类中的每个测试之前执行，以便执行某些必要的先决条件
@BeforeClass	作用于静态方法，表示其在类的所有测试之前必须执行一次。该注解一般用于提供测试计算、共享配制方法（如数据库连接）
@After	作用于方法，表示该方法在每项测试后执行（如执行每一个测试后重置某些变量，删除临时变量等）
@AfterClass	作用于静态方法，表示该方法会在所有测试用例之后执行，AlterClass 注解适用于声明方法去清理一些资源（如数据库连接）
@Ignore	作用于方法，当想要暂时禁用特定的测试执行时，可以使用这个注解，每个被注解为@Ignore 的方法将不再执行
@Runwith	作用于类，用来确定该类运行所依赖的运行期。也可以不标注，会使用默认运行器
@Parameters	用于使用参数化功能
@SuiteClasses	用于套件测试

着手使用 JUnit 为程序做单元测试之前，需要添加相关依赖。因为目的是为 Spring Boot 程序做单元测试，使用 Spring 提供的 spring-boot-starter-test 即可，其中包含 JUnit 依赖：

```xml
<dependency>
    <groupId>org.springframework.boot</groupId>
    <artifactId>spring-boot-starter-test</artifactId>
    <scope>test</scope>
</dependency>
```

1. 基础示例

【示例 6-1】

首先可以通过一个简短的例子了解部分注解的使用方式，/test/java/com/example/unit/QuickBeforeAfterTest.java：

> **注 意**
>
> 测试相关的 Java 文件基本都需要放置于/test/java 目录下。

```java
import org.junit.*;

public class QuickBeforeAfterTest {

    @BeforeClass
    public static void beforeClass(){
        //在整个类所有测试方法前执行的静态方法（仅执行一次）
        System.out.println("Before Class");
    }

    @Before
    public void setup(){
```

```java
        //在所有测试方法之前执行（执行数次）
        System.out.println("Before Test");
    }

    @Test
    public void test1() {
        //测试方法 1
        System.out.println("test1 executed");
    }

    @Test
    public void test2() {
        //测试方法 2
        System.out.println("test2 executed");
    }

    @After
    public void teardown() {
        //在所有测试方法之后执行（执行数次）
        System.out.println("After test");
    }

    @AfterClass
    public static void afterClass(){
        //在整个类所有测试方法后执行的静态方法（仅执行一次）
        System.out.println("After Class");
    }
}
```

执行该组测试用例可以对各注解标注方法的执行顺序有一个简单的认识，以上示例的执行顺序如下：

- @BeforeClass 标注的 beforeClass()方法。
- @Before 标注的 setup()方法。
- @Test 标注的 test1()方法。
- @After 标注的 teardown()方法。
- @Before 标注的 setup()方法。
- @Test 标注的 test2()方法。
- @After 标注的 teardown()方法。
- @AfterClass 标注的 afterClass()方法。

2. 异常处理

@Test 用于标注测试用例之外，还可以指定测试过程中预期遇到的异常。使用@Test 以及其他两种对于异常的测试如下：

```java
@Test(expected = FileNotFoundException.class)
public void testReadFile() throws IOException {
    //文件名为一个不存在的文件
    FileReader reader = new FileReader("non-existent.txt");
    reader.read();
```

```java
        reader.close();
}

@Test
public void testReadFile2() {
    try {
        FileReader reader = new FileReader("test.txt");
        reader.read();
        reader.close();
        fail("Expected an IOException to be thrown");
    } catch (IOException e) {
        //并不推荐该方法，e.getMessage()的内容不能保证是固定的
        assertThat(e.getMessage(), is("test.txt （系统找不到指定的文件。)"));
    }
}

@Rule
public ExpectedException thrown = ExpectedException.none();

@Test
public void testReadFile3() throws IOException {
    thrown.expect(IOException.class);
    //同样不推荐该方法，原因同上。仅供参考
    thrown.expectMessage(startsWith("test.txt （系统找不到指定的文件。)"));
    FileReader reader = new FileReader("test.txt");
    reader.read();
    reader.close();
}
```

其中第三个示例中用到了@Rule 注解。该注解结合 org.junit.rules.TestRule 接口的实现类，可以非常灵活地添加或重新定义测试类中每个测试方法的行为。

3. 参数化测试

自定义运行器 Parameterized 结合@Parameters 可以将测试参数化，根据提供的多组测试数据运行多个测试用例。

例如当前有待测类，用于计算斐波那契数列的 Fibonacci.java：

```java
public class Fibonacci {
    public static int compute(int n) {
        int result = 0;
        if (n <= 1) {
            result = n;
        } else {
            result = compute(n - 1) + compute(n - 2);
        }
        return result;
    }
}
```

对应的测试代码如下：

```java
@RunWith(Parameterized.class)
public class FibonacciTest {
```

```
    @Parameters
    public static Collection<Object[]> data() {
        return Arrays.asList(new Object[][] {
                { 0, 0 }, { 1, 1 }, { 2, 1 }, { 3, 2 }, { 4, 3 }, { 5, 5 }, { 6, 8 }
        });
    }

    private final int fInput;

    private final int fExpected;

    public FibonacciTest(int input, int expected) {
        this.fInput = input;
        this.fExpected = expected;
    }

    @Test
    public void test() {
        assertEquals(fExpected, Fibonacci.compute(fInput));
    }
}
```

注入数据的方式除了声明测试类的构造器之外，还可以使用@Parameterized.Parameter 实现字段注入，示例代码如下：

```
@RunWith(Parameterized.class)
@RunWith(Parameterized.class)
public class FibonacciTest {
    @Parameters
    public static Collection<Object[]> data() {
        return Arrays.asList(new Object[][]{
                {0, 0}, {1, 1}, {2, 1}, {3, 2}, {4, 3}, {5, 5}, {6, 8}
        });
    }

    /*必须使用public，下同*/
    @Parameterized.Parameter
    public int fInput;

    @Parameterized.Parameter(1)
    public int fExpected;

    @Test
    public void test() {
        assertEquals(fExpected, Fibonacci.compute(fInput));
    }
}
```

4. 使用套件汇总多个测试

通过 Suite 运行器配合@Suite.SuiteClasses 注解，可以实现多个测试类同时运行，示例代码 TestSuite.java 如下：

```
@RunWith(Suite.class)
```

```
@Suite.SuiteClasses(
        {QuickBeforeAfterTest.class,
               FibonacciTest.class}
)
public class TestSuite {
    //类的内容为空
    //该类仅作为以上注解的容器（持有者）而存在
}
```

运行该测试类等同于运行完 QuickBeforeAfterTest 后接着运行 FibonacciTest。

5. 忽略测试用例

如果遇到一些原因想要忽略测试用例，则可以通过@Ignore 注解完成。@Ignore 提供了一个可选参数用于记录忽略测试用例的理由。

```
@Ignore("结果显而易见")
@Test
public void testSame() {
    assertThat(1, is(1));
}
```

6.1.3 JUnit 5 简介

JUnit 5 是 JUnit 的最新版本。有别于之前的版本，JUnit 5 由三个不同子项目的模块组成。

JUnit 5 = JUnit 平台+ JUnit Jupiter + JUnit Vintage

JUnit 平台，其主要作用是在 JVM 上启动测试框架。它使用了一个抽象的 TestEngineAPI 来定义运行在平台上的测试框架，同时还支持通过命令行、Gradle 和 Maven 来运行平台。

JUnit Jupiter，包含了 Junit 5 最新的编程模型和扩展机制。

JUnit Vintage，允许在平台上运行 JUnit 3 和 JUnit 4 的测试用例。

熟悉 JUnit 5 也可以从 JUnit 5 里的新注解开始，这部分均来自于 JUnit Jupiter，参见表 6.2。

表 6.2 JUnit 5 的新注解

注解	描述
@Test	表示方法是测试方法。与 JUnit 4 的@Test 注解不同，此注解不声明任何属性，因为 JUnit Jupiter 中的测试扩展基于其自己的专用注解进行操作。除非重写这些方法，否则它们将被继承
@ParameterizedTest	表示方法是参数化测试。除非重写这些方法，否则它们将被继承
@RepeatedTest	表示方法是重复测试的测试模板。除非重写这些方法，否则它们将被继承
@TestFactory	表示方法是用于动态测试的测试工厂。除非重写这些方法，否则它们将被继承
@TestTemplate	表示方法是测试用例的模板，设计用于根据已注册提供者返回的调用上下文的数量被多次调用的测试用例。除非重写这些方法，否则它们将被继承
@TestMethodOrder	用于为带注解的测试类配置测试方法的执行顺序；类似于 JUnit 4 的 @FixMethodOrder。这样的注解是继承的
@TestInstance	用于为带注解的测试类配置测试实例生命周期。这样的注解是继承的

（续表）

注解	描述
@DisplayName	声明测试类或测试方法的自定义显示名称。这样的注解不是继承的
@DisplayNameGeneration	声明测试类的自定义显示名称生成器。这样的注解是继承的
@BeforeEach	表示该注解的方法应该在 @Test、@RepeatedTest、@ParameterizedTest 或 @TestFactory 方法之前执行；类似于 JUnit 4 的@Before。除非重写这些方法，否则它们将被继承
@AfterEach	表示该注解的方法应该在 @Test、@RepeatedTest、@ParameterizedTest 或 @TestFactory 方法之后执行；类似于 JUnit 4 的@After。除非重写这些方法，否则它们将被继承
@BeforeAll	表示该注解的方法应该在 @Test、@RepeatedTest、@ParameterizedTest 和 @TestFactory 方法之前执行；类似于 JUnit 4 的@BeforeClass。此类方法是继承的（除非它们被隐藏或覆盖），并且必须被继承（除非 static 使用"每类"测试实例生命周期）
@AfterAll	表示该注解的方法应该在 @Test、@RepeatedTest、@ParameterizedTest 和 @TestFactory 方法之后执行；类似于 JUnit 4 的@AfterClass。此类方法是继承的（除非它们被隐藏或覆盖），并且必须被继承（除非 static 使用"每类"测试实例生命周期）
@Nested	表示带注解的类是一个非静态的嵌套测试类。@BeforeAll 和@AfterAll 方法不能直接使用@Nested 测试类，除非"每级"测试实例的生命周期被使用。这样的注解不是继承的
@Tag	用于在类或方法级别上声明用于过滤测试的标签；类似于 TestNG 中的测试组或 JUnit 4 中的类别。此类注解在类级别继承，而不在方法级别继承
@Disabled	用于禁用测试类或测试方法；类似于 JUnit 4 的@Ignore。这样的注解不是继承的
@Timeout	如果执行超过给定的持续时间，则使测试、测试工厂、测试模板或生命周期方法失败。这样的注解是继承的
@ExtendWith	用于声明性地注册扩展。这样的注解是继承的
@RegisterExtension	用于通过字段以编程方式注册扩展。除非被遮盖，否则这些字段将被继承
@TempDir	用于通过生命周期方法或测试方法中的字段注入或参数注入来提供临时目录；位于 org.junit.jupiter.api.io 包装中

1. JUnit 4 与 JUnit 5 的区别

使用 JUnit 4 与 JUnit 5 大体上相同，但有几个不同之处首先需要注意。在导入的包方面，JUnit 5 使用新的 org.junit.jupiter 包。例如，org.junit.junit.Test 变成了 org.junit.jupite-r.api.Test。

另外部分关键注解的使用也发生了变化。

@Test 注解不再有参数，参数被移至函数中。例如 6.1.2 小节中预测抛出异常的写法，在 JUnit 5 中使用 Lambda 的方法予以实现：

```
@Test
public void testReadFile() throws IOException {
    Assertions.assertThrows(Exception.class, () -> {
        //文件名为一个不存在的文件
        FileReader reader = new FileReader("non-existent.txt");
```

```
        reader.read();
        reader.close();
    });
}
```

另外，还可以实现超时。JUnit 4 中的超时：

```
@Test(timeout = 10)
public void testFailWithTimeout() throws InterruptedException {
    Thread.sleep(100);
}
```

在 JUnit 5 中：

```
@Test
public void testFailWithTimeout() throws InterruptedException {
    Assertions.assertTimeout(Duration.ofMillis(10), () -> Thread.sleep(100));
}
```

另外是一些名词上的变化：

- @Before 变成了@BeforeEach。
- @After 变成了@AfterEach。
- @BeforeClass 变成了@BeforeAll。
- @AfterClass 变成了@AfterAll。
- @Ignore 变成了@Disabled。
- @Category 变成了@Tag。
- @Rule 和@ClassRule 没有了，用@ExtendWith 和@RegisterExtension 代替。

2. 从 JUnit 4 迁移

因为 JUnit 5 中包含 JUnit Vintage 引擎,该引擎可以支持在 JUnit Platform 之上执行基于 JUnit 3 和 JUnit 4 的现有测试。因此当前情况下，同时维护 JUnit 4 和基于 JUnit Jupiter 的测试不会有问题。不过，依然建议使用基于 JUnit Jupiter 或者说 JUnit 5 的功能开发测试用例，理由如下：

- JUnit 5 利用了 Java 8 或更高版本的特性，例如上文提到的 Lambda 函数。
- 为描述、组织和执行测试提供了很多新功能，例如可以使用@DisplayName 定义测试类和测试方法的名称。
- 被组织成多个库，运行将仅需要的功能导入到项目中。
- 可以同时使用多个扩展，这是 JUnit 4 无法做到的（一次只能使用一个 runner）。

6.2 断　言

在编写代码的过程中，往往需要做出一些判断。断言的作用在于在代码中捕捉这些判断，通过将判断的内容以布尔表达式（Boolean Expression）的形式展现，而后当表达式的运行结果为 false 时，终止程序。这种方式可以被当作是异常处理的一种形式。Java 在 JDK 1.4 中通过关键字 assert

提供了该手段，这也是最为常见的断言。本节将介绍 Spring Boot 的自动化测试中，使用断言的方式与技巧，从而编写出可靠的测试用例。

6.2.1 assert 关键字

JDK 中自带的 assert 关键字是最基础的断言方式，使用方式很简单，分为两种：
- assert <布尔表达式>。
- assert <布尔表达式>:<错误信息表达式>。

这两种方式都是对布尔表达式进行判断，如果表达式结果为 false，则抛出错误 java.lang.AssertionError 以终止程序。两者不同的是，第二种方式会另外在抛出的错误信息中将错误信息表达式的内容也附加进去。示例如下：

```
@Test
void assertTrue_ThenAssertFalseWithInfo() {
    //断言 1 结果为 true，则继续往下执行
    assert 1 == 1;
    System.out.println("第一个断言顺利通过");
    //断言 2 结果为 false，抛异常程序终止
    assert 1 > 1 : "停下，这里是错误信息";
    System.out.println("第二个断言顺利通过");
}
```

6.2.2 JUnit 4 里的断言

JUnit 4 为了帮助开发人员编写测试用例提供了一组静态的断言方法，可以直接使用例如：Assert.assertEquals(...)或者导入包 import static org.junit.Assert.*;之后再对其中方法进行调用。常用的方法有：

- assertEquals(expected, actual)：断言两个对象相等，等于 assert expected.equals(actual)。
- assertNotEquals(first, second)：断言两个对象不相等。
- assertNull(object)：断言该对象为 Null。
- assertNotNull(object)：断言该对象不为 Null。
- assertSame(expected, actual)：断言两个引用是否指向同一对象，视同 assert expected == actual。
- assertNotSame(unexpected, actual)：断言两个引用是否指向不同对象。
- assertTrue(String message, boolean condition)：断言 condition 结果为 True，否则将 message 当作错误信息一起抛出。
- assertFalse(String message, boolean condition)：断言 condition 结果为 False，否则将 message 当做错误信息一起抛出。
- assertArrayEquals(String message, Object[] expecteds, Object[] actuals)：断言数组相等。
- fail：使测试立刻失败。

使用以上方法来代替 assert 关键字，可以提高测试用例的代码可读性。这是一份易维护的代码

所需要具备的。

6.2.3　assertThat 方法

JUnit 4 中虽然已经提供了一批方法帮助开发人员优化测试代码,但是可以观察到 JUnit 4 中自带的基础断言方法并不足以覆盖测试所需的方方面面。因此,JUnit 4 另外集成了 Hamcrest,提供了 assertThat(T actual, Matcher<? super T> matcher)方法。其中 Matcher 是 Hamcrest 提供的匹配符,这些匹配符非常全面并且命名接近自然语言。这一方式可以更好地完成优化断言表达的工作。为展示 assertThat 具有很强可读性,列举示例如下:

```
@Test
void assertThatTest() {
    //断言 1 等于 1
    assertThat(1, equalTo(1));
    //50 是否大于 30 并且小于 60
    assertThat("错误信息", 50, allOf(greaterThan(30), lessThan(60)));
    //判断字符串是否以.txt 结尾
    assertThat("错误信息", "abc.txt", endsWith(".txt"));
}
```

从示例中可以看出,阅读 assertThat 所编写的断言语句,如同在读一句话。

【示例 6-2】

常用的方法示例如下:

```
@Test
public void showMatchers() {
    //equalTo 断言输入项逻辑上等于匹配项
    assertThat(1, equalTo(1));

    //hasToString 断言输入项调用.toString 方法后与匹配项相等
    assertThat(1, hasToString("1"));

    //is 语法糖单独使用等同于 equalTo
    assertThat(1, is(1));

    //结合其他匹配器使用则相当于直接调用其他匹配器,仅用作提高可读性
    assertThat(1, is(hasToString("1")));

    //closeTo 输入项是否接近于匹配项,例如 1.03 处于 1.06+/- 0.3,即 1.03 到 1.09 这
范围内
    assertThat(new BigDecimal("1.03"), closeTo(new BigDecimal("1.06"),
new BigDecimal("0.03")));

    //sameInstance 判断输入项与匹配项是否为同一实例
    assertThat(BigDecimal.ONE, sameInstance(BigDecimal.ONE));

    //greaterThan 断言输入项大于匹配项,需要都是 Comparable 类型的对象
    //类似的还有 greaterThanOrEqualTo, lessThan, lessThanOrEqualTo
    assertThat(new BigDecimal("1.03"), greaterThan(BigDecimal.ONE));
```

```java
    //hasProperty 判断javaBean是否包含匹配项声明的属性名
    assertThat(new User(), hasProperty("username"));

    //array 将数组的内容拿出来依次作比较
    assertThat(new Integer[]{1, 2, 3}, array(is(1), is(2), is(3)));

    //hasItemInArray 断言数组内包含某元素
    assertThat(new Integer[]{1, 2, 3}, hasItemInArray(2));

    List<Integer> list = new ArrayList<>();
    list.add(1);
    list.add(2);
    //hasItem 断言集合中包含Item
    assertThat(list, hasItem(1));

    Map<Integer, String> map = new HashMap<>();
    map.put(1, "1");
    map.put(2, "2");
    //hasEntry 判断map中是否包含entry
    assertThat(map, hasEntry(1, "1"));

    //equalToIgnoringCase 在忽略大小写的情况下断言两字符串相等
    assertThat("Process finished with exit code 0",
equalToIgnoringCase("process finished with exit code 0"));

    //equalToCompressingWhiteSpace 在忽略空格的情况下断言两字符串相等
    assertThat("Process finished with exit code 0",
equalToCompressingWhiteSpace("Process  finished  with exit code 0"));

    //containsString 判断输入项内包含字符串, 类似的还有endsWith, startsWith
    assertThat("Process finished with exit code 0", containsString("0"));

    //allOf 组合匹配器, 如果所有匹配器都匹配才匹配, 类似于&&
    assertThat(1, allOf(greaterThan(0), lessThan(2)));

    //any 组合匹配器, 如果任意匹配器匹配就匹配, 类似于||
    assertThat(1, anyOf(greaterThan(0), lessThan(2)));

    //not 组合匹配器, 非类似!
    assertThat(1, not(greaterThan(2)));
}
```

6.2.4 自定义 Hamcrest 匹配器

虽然 Hamcrest 内置了非常多的匹配器供开发人员使用, 但总会存在不能精确满足需求的时候。这时候就不得不通过自定义匹配器的方式来实现了。自定义 Hamcrest 匹配器的方式分两种: 实现 BaseMatcher<T>或者 TypeSafeMatcher<T>。

1. 实现 BaseMathcer<T>

以实现一个判断字符串的自定义匹配器为例, 示例如下:

```java
public class IsHelloWorldMatcher extends BaseMatcher<String> {
```

```java
    public static <T> Matcher<String> isHelloWorld() {
        //用以返回一个匹配器示例
        return new IsHelloWorldMatcher();
    }

    @Override
    public boolean matches(Object actual) {
        //匹配业务逻辑
        if (actual == null) {
            return false;
        }
        String s = (String) actual;
        return s.startsWith("Hello") && s.endsWith("World");
    }

    @Override
    public void describeTo(Description description) {
        //对于输入项的描述,在断言失败后将会被打印到 console 中
        description.appendText("a string that start with \"Hello\" and end with \"World\"");
    }
}
```

为了保证风格与原生匹配器一致,首先需要定义一个静态方法返回当前类的示例。然后便是 matches(Object actual)用于表示匹配逻辑。最后 describeTo(Description description)定义期望输入项的描述。调用示例:

```java
@Test
public void helloWorldCustomTest() {
    assertThat("Hey Java World", isHelloWorld());
}
```

该调用示例将会因为断言失败而抛出 AssertionError 异常,异常内容如图 6.2 所示。

图 6.2　自定义匹配器断言异常内容

2. 实现 TypeSafeMatcher<T>

实现 TypeSafeMatcher 与 BaseMathcer 非常类似,只不过 TypeSafeMatcher 的匹配逻辑会将输入项做 null 处理并进行类型转换。例如下例将会是一个永不会匹配成功的匹配器:

```java
public class NeverSuccess extends TypeSafeMatcher<String> {
    public static <T> Matcher<String> neverSuccess() {
        return new NeverSuccess();
    }
```

```
    @Override
    protected boolean matchesSafely(String item) {
        return item == null;
    }

    @Override
    public void describeTo(Description description) {
        description.appendText("never success……");
    }
}
```

调用示例：

```
@Test
public void neverSuccessTest() {
    assertThat(null, neverSuccess());
}
```

调用结果如图 6.3 所示。

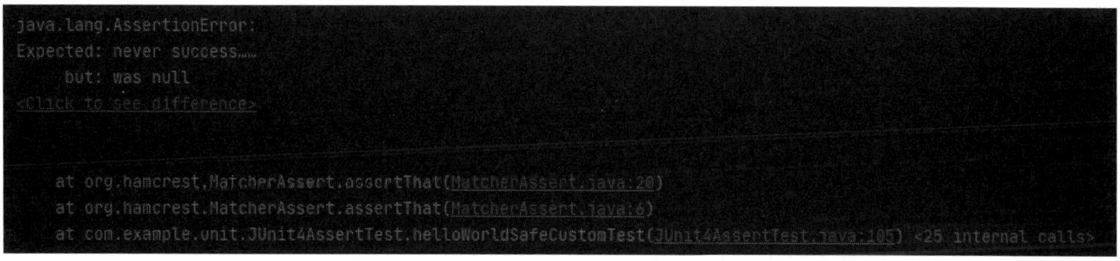

图 6.3　neverSuccess()匹配结果

6.2.5　断言框架 AssertJ

AssertJ 和 Hamcrest 一样是用于断言的框架。虽然 AssertJ 没有 Hamcrest 那样知名，但在 JUnit 5 中 AssertJ 已经代替 Hamcrest 成为了推荐选项。

1．流式断言

AssertJ 的语法为流式（Fluent），所有的断言以 assertThat(value)开始，value 中为需要进行匹配的对象。当断言的匹配项很多时，这种语法的使用将会很方便。示例如下：

```
@Test
public void fluentTest() {
    Integer[] arr = new Integer[]{1, 2, 3};

    assertThat(1)
            .isEqualTo(1)
            .isLessThan(2)
            .isIn(arr);

    assertThat(arr)
            .hasSize(3)
            .contains(2, 3)
```

```
            .doesNotContain(4);

    assertThat("assert that")
            .startsWith("assert")
            .endsWith("that");
}
```

2. 对象间的比较

AssertJ 提供了一些针对对象字段比较的方法，即便是非常复杂的对象，也可以使用这些方法轻松完成断言比较。例如有两种 POJO 类，User 和 Visitor：

```
@Accessors(chain = true)
@Setter
@Getter
public class User {
    private String username;
    private String firstName;
    private String lastName;
}

@Accessors(chain = true)
@Setter
@Getter
public class Visitor {
    private String username;
}
```

即便是所比较的对象类型不同，也可以拿来当作匹配项进行比较。示例代码如下：

```
@Test
public void compareObjects() {
    User user1 = new User()
            .setUsername("anonymous")
            .setFirstName("zhang")
            .setLastName("san");
    User user2 = new User()
            .setUsername("anonymous")
            .setFirstName("zhang")
            .setLastName("san");
    User user3 = new User()
            .setUsername("anonymous")
            .setFirstName("li")
            .setLastName("si");
    Visitor visitor = new Visitor()
            .setUsername("anonymous");
    assertThat(user1)
            //各字段均需要匹配
            .isEqualToComparingFieldByField(user2)
            //忽略给出的字段
            .isEqualToIgnoringGivenFields(user3, "firstName", "lastName")
            //匹配给出的字段
            .isEqualToComparingOnlyGivenFields(visitor, "username");
}
```

使用 AssertJ 可以编写出更加容易理解的测试用例，增加测试的价值并且降低了编写成本。这不仅起到了鼓励开发人员编写更多自动化测试用例的作用，另一方面也方便项目维护人员理解测试用例以及所测试的功能。

6.3 测试中的模拟行为 Mock

Mock，可以理解为测试替身，也有部分人使用 Test Doubles 来描述。这一概念指的是在测试过程中，对于一些不易获取或者构造的对象，通过创造一个替身对象来模拟这些不易获取对象的行为。本节将介绍在 Spring Boot 程序的开发过程中，如何使用 Mock 来更好地创建测试用例。

6.3.1 测试替身

当项目的规模逐渐变得庞大，类与类之间，模块与模块之间存在着难以分割的依赖与耦合。处于这种情况之下，想要进行细小粒度的单元测试会变得异常棘手。例如，构造一个 A 类实例所需要的前置依赖有 B、C、D、E 这三个类，如图 6.4 所示。

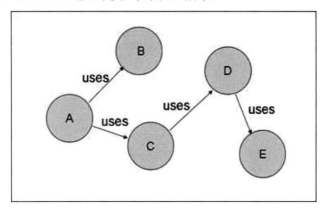

图 6.4 复杂的依赖关系

要创建 A 类的实例之前又需要创建另外的实例。如果这些类还有前置依赖，将会导致测试的过程变得异常复杂。解决该类型的问题，就可以选择使用 Mock。

测试替身分为不同种类，各自有不同的作用。它们分别是：

- Dummy Object
- Stub
- Spy
- Mock Object

1. Dummy Object

Dummy Object 可以被称作虚拟对象，或者占位符。该测试替身的作用在于填充无关参数，以保证测试用例可以成功运行。实际使用中，Dummy Object 或许就是一个"null"。

> **注　意**
>
> Dummy Object 所替代的部分在测试过程中是不允许被调用的。

2. Stub

在许多情况下，SUT（被测系统，System Under Test）所运行的环境或者上下文会极大地影响 SUT 的行为。因为 SUT 之外的组件存在不可控或者并不方便加入到测试的情况，为了获得对 SUT "间接输入"足够好的控制，可能需要使用可供开发人员支配的内容来替换这些上下文。Stub（测试桩）便是完成这一目的的手段，它可以根据预期设置返回值，并代替对应组件在逻辑中返回。通过这种方式，Stub 就能帮助开发人员将 SUT 中的逻辑与外部组件解耦。

3. Spy

同样在许多情况下，"SUT 所运行的环境或者上下文会极大地影响 SUT 的行为"这一原因。为了获取对 SUT "间接输出"足够好的可见性，需要捕获一些输出的上下文来代替。Spy 是其中相对简单直观的一种方式，它公开了 SUT 的间接输出，以便对其进行验证。

4. Mock Object

Mock Object（模拟对象）用以实现一个功能强大的行为验证。与 Spy 在某种程度上很相似，它的主要作用也是在于获取对 SUT "间接输出"的可见性，但是使用细节方面与 Spy 存在差异。比如 Mock Object 在创建时必须根据预定义的方法调用来返回结果。

6.3.2　Mockito 框架

Mockito 是 Java 开发中常用的 Mock 工具。因为 Spring Boot Test 中已经包含 Mockito 的依赖，所以不需要再另外引入即可使用。在代码中的使用方式非常简单。

【示例 6-3】

首先展示一个简单的示例：

```
//推荐静态导入，可以使代码更简洁
import static org.mockito.Mockito.*;

// 创建 Mock 对象
List mockedList = mock(List.class);

// 使用 Mock 对象
mockedList.add("one");
mockedList.clear();

// 验证目标方法已被调用
verify(mockedList).add("one");
verify(mockedList).clear();
```

一旦 Mock Object 被创建，它将会记录它自身方法的调用情况。之后可以通过 verify 方法对其进行验证。

1. Stubbing

Mockito 中的 Stub 通过对 Mock 对象进行 Stubbing（打测试桩）来实现。示例代码如下：

```
//我们可以 Mock 具体的类型，不仅只是接口
LinkedList mockedList = mock(LinkedList.class);

//测试桩
when(mockedList.get(0)).thenReturn("first");
when(mockedList.get(1)).thenThrow(new RuntimeException());

//输出 first
System.out.println(mockedList.get(0));

//调用该方法将抛出异常
//System.out.println(mockedList.get(1));

//因为 get(999) 没有打测试桩，因此输出 Null
System.out.println(mockedList.get(999));

//验证 get(0)被调用的次数
verify(mockedList).get(0);
```

2. @Mock、@Spy、@Captor 和@InjectMocks

通过使用 Mockito 注解，可以方便地实现 Mock Object 的创建与注入，而不必再调用 Mockito.mock 方法。不使用注解的示例如下：

```
@Test
public void whenNotUseMockAnnotation_thenCorrect() {
    List mockList = Mockito.mock(ArrayList.class);

    mockList.add("one");
    verify(mockList).add("one");
    assertThat(mockList.size())
            .isEqualTo(0);

    when(mockList.size()).thenReturn(100);
    assertThat(mockList.size())
            .isEqualTo(100);
}
```

使用注解的示例如下：

```
@Mock
List<String> mockedList;

@BeforeEach
public void init() {
    MockitoAnnotations.initMocks(this);
}

@Test
public void whenUseMockAnnotation_thenMockIsInjected() {
    mockedList.add("one");
    verify(mockedList).add("one");
```

```
    assertThat(mockedList.size())
            .isEqualTo(0);

    when(mockedList.size()).thenReturn(100);
    assertThat(mockedList.size())
            .isEqualTo(100);
}
```

通过@Spy 的 Spy 同样可以使用 Stubbing 的操作设定方法的返回值,但如果没有对其 Stubbing,则最终方法会委托给原本的实例来执行。示例代码如下:

```
@Spy
List<String> spiedList = new ArrayList<>();

@Test
public void whenUseSpyAnnotation_thenSpyIsInjectedCorrectly() {
    spiedList.add("one");
    spiedList.add("two");

    verify(spiedList).add("one");
    verify(spiedList).add("two");

    assertThat(spiedList.size())
            .isEqualTo(2);

    //stubbing
    doReturn(100).when(spiedList).size();
    assertThat(spiedList.size())
            .isEqualTo(100);
}
```

注解@Captor 用于帮助创建 ArgumentCaptor 对象。ArgumentCaptor 与 verify 方法配合使用,用于捕获方法参数。示例代码如下:

```
@Captor
ArgumentCaptor argument;

@Test
public void argumentCaptorTest() {
    List list1 = mock(List.class);
    List list2 = mock(List.class);
    list1.add("Scott");
    list2.add("Brian");
    list2.add("Jim");

    verify(list1).add(argument.capture());
    assertThat(argument.getValue())
            .isEqualTo("Scott");

    verify(list2, times(2)).add(argument.capture());

    assertThat(argument.getValue())
            .isEqualTo("Jim");
    assertThat(argument.getAllValues().toArray())
            .isEqualTo(new Object[]{"Scott", "Brian", "Jim"});
```

@InjectMocks 用于创建一个示例,并且将使用了@Mock 的字段注入其中。示例代码如下,首先有一个 MyDictionary 类:

```java
public class MyDictionary {
   Map<String, String> wordMap;

   public MyDictionary() {
      wordMap = new HashMap<String, String>();
   }
   public void add(final String word, final String meaning) {
      wordMap.put(word, meaning);
   }
   public String getMeaning(final String word) {
      return wordMap.get(word);
   }
}
```

在使用时,将@InjectMocks 结合@Mock 使用:

```java
@Mock
Map<String, String> wordMap;

@InjectMocks
MyDictionary dic = new MyDictionary();

@Test
public void whenUseInjectMocksAnnotation_thenCorrect() {
   when(wordMap.get("word")).thenReturn("meaning");

   assertThat(dic.getMeaning("word"))
         .isEqualTo("meaning");
}
```

6.4 集成测试

在完成对单元测试以及相关测试技巧的介绍之后,本节将介绍基于 Spring Boot Test 的集成测试(Integration Test)。在基于 Spring Boot 的开发当中,单元测试与集成测试基本上以是否需要启动 Spring 来区分。在 Spring 官方文档的描述中,单元测试的运行非常快,因为单元测试并不依赖其他外部组件或框架。集成测试则是依赖 Spring 框架才可以正常进行的测试。为了帮助开发人员在集成测试中只测试被关注的功能点,Spring Boot Test 提供了若干注解。

6.4.1 @WebMvcTest 注解

@WebMvcTest 注解用于帮助开发人员开展侧重点在展示层面的集成测试,它作用且仅可作用于 Spring MVC 组件。使用该注解之后,全局的自动配置将被禁用,取而代之的是启用 MVC 测试相关的配置。

【示例 6-4】

待测对象 UserController：

```java
@RestController
@RequiredArgsConstructor
@RequestMapping("/api/user")
public class UserController {

    private final UserService userService;

    @GetMapping("/all")
    public Iterable<User> findAll() {
        return userService.findAll();
    }

    @GetMapping("/{login:[\\D]+}")
    public User findOne(@PathVariable String login) {
        //login 只匹配非数字的参数
        //单个参数绑定单个入参
        User result = userService.findByLogin(login);
        if (result == null) {
            throw new ResponseStatusException(HttpStatus.NOT_FOUND, "This user does not exit");
        }
        return result;
    }

    @GetMapping(value = "/{firstName}/{lastName}")
    public User findOne(@PathVariable String firstName, @PathVariable String lastName) {
        //多个参数绑定多个入参
        User result = userService.findByName(firstName, lastName);
        if (result == null) {
            throw new ResponseStatusException(HttpStatus.NOT_FOUND, "This user does not exit");
        }
        return result;
    }

    @PostMapping("/one")
    public User createOne(@RequestBody @Valid User newOne) {
        return userService.save(newOne);
    }
}
```

@WebMvcTest 集成测试基础结构：

```java
@WebMvcTest(controllers = UserController.class)
public class UserControllerTest {

    @Autowired
    private MockMvc mockMvc;
```

```
    @Autowired
    private ObjectMapper objectMapper;

    @MockBean
    private UserService userService;

    @Test
    void whenValidInput_thenReturns200() throws Exception {
        mockMvc.perform(...);
    }
}
```

示例中，使用@WebMvcTest 指定需要测试的控制器。属性中的 MockMvc 为 Spring-Test 提供的测试工具。MockMvc 实现了对 HTTP 请求的模拟，能够直接使用网络的形式转换到 Controller 的调用。这样的实现方式优点在于测试速度快并且不依赖网络环境。

ObjectMapper 用于测试过程中对于 JSON 与对象间的转换，也可以使用其他的依赖来实现，比如 GSON。

@MockBean 为 Mockito 提供，用于模拟测试对象的依赖。该注解可以帮助开发人员调整测试的粒度。如果仅关注 UserController 中的逻辑，则需要@MockBean 参与其中，以实现在测试程序运行中对其依赖的屏蔽。

1. 验证 HTTP 请求匹配

验证控制器对某个 HTTP 请求的监听非常简单，调用 MockMvc 的 perform()方法并提供想要测试的 URL 即可：

```
@Test
void whenGetApiUser_thenReturns200() throws Exception {
    mockMvc.perform(MockMvcRequestBuilders.get("/api/user/all"))
            .andExpect(status().isOk());
}
```

2. 验证输入

为了验证输入内容是否可以成功序列化，因此必须在测试过程中在请求体中附上请求内容：

```
@Test
void whenPostApiUserOne_thenSuccess() throws Exception {
    User user = new User()
            .setLogin("someone")
            .setFirstName("sun")
            .setLastName("wan");
    mockMvc.perform(MockMvcRequestBuilders.post("/api/user/one")
            .contentType("application/json")
            .content(objectMapper.writeValueAsString(user)))
            .andExpect(status().isOk());
}
```

3. 验证输入检查

测试对象中已使用@Valid 开启了输入检查，并且 User 中也已经开启了对应的输入检查注解：

```
@Accessors(chain = true)
@Setter
```

```java
@Getter
@Entity
@EqualsAndHashCode
public class User {
    @Id
    @GeneratedValue
    private Long id;
    @NotNull
    private String login;
    @NotNull
    private String firstName;
    @NotNull
    private String lastName;
    private String description;
}
```

根据配置构造一个会引发 400 请求错误的请求内容：

```java
@Test
void whenPostApiUserOne_thenFail() throws Exception {
    User user = new User()
            .setLogin(null)
            .setFirstName("sun")
            .setLastName("wan");
    mockMvc.perform(MockMvcRequestBuilders.post("/api/user/one")
            .contentType("application/json")
            .content(objectMapper.writeValueAsString(user)))
            .andExpect(status().isBadRequest());
}
```

4. 验证依赖项调用情况（业务逻辑）

示例中测试对象里的业务逻辑因为是用于演示而编写得非常简单，实际开发中的逻辑或许会相对复杂许多。因此，从控制器输入然后到调用依赖的过程中，或许会改变输入内容。通过 Mockito 提供的功能，可用以确认这个过程中测试对象对依赖的调用情况：

```java
@Test
void whenPostApiUserOne_thenCheckSave() throws Exception {
    User user = new User()
            .setLogin("someone")
            .setFirstName("sun")
            .setLastName("wan");
    mockMvc.perform(MockMvcRequestBuilders.post("/api/user/one")
            .contentType("application/json")
            .content(objectMapper.writeValueAsString(user)));

    //创建参数捕获器
    ArgumentCaptor<User> userArgumentCaptor =
ArgumentCaptor.forClass(User.class);
    //确认调用次数并且设置参数捕获器
    verify(userService, times(1)).save(userArgumentCaptor.capture());
    //确认输入参数
    assertThat(userArgumentCaptor.getValue().getLogin())
            .isEqualTo("someone");
    assertThat(userArgumentCaptor.getValue().getFirstName())
```

```
            .isEqualTo("sun");
    assertThat(userArgumentCaptor.getValue().getLastName())
            .isEqualTo("wan");
}
```

5. 验证输出

在对业务逻辑进行验证之后，可能还需要对于输出内容进行验证。具体方式是通过 MockMvc 的 andReturn 获取 MvcResult，然后对 MvcResult 中的响应体进行验证：

```
@Test
void whenPostApiUserOne_thenCheckResponse() throws Exception {
    User user = new User()
            .setLogin("someone")
            .setFirstName("sun")
            .setLastName("wan");
    //给 userService 打测试桩
    when(userService.save(eq(user))).thenReturn(user);
    MvcResult result =
mockMvc.perform(MockMvcRequestBuilders.post("/api/user/one")
            .contentType("application/json")
            .content(objectMapper.writeValueAsString(user)))
            .andExpect(status().isOk())
            .andReturn();
    assertThat(result.getResponse().getContentAsString())
            .isEqualToIgnoringCase(objectMapper.writeValueAsString(user))
;
}
```

6.4.2 @DataJpaTest 注解

@DataJpaTest 与 @WebMvc 类似，它用于帮助开发人员开展侧重于存储层的集成测试，它作用于且仅可作用于 JPA 组件。使用该注解之后，全局的自动配置将被禁用，取而代之的是启用 JPA 测试相关的配置。

【示例 6-5】

示例代码如下：

```
@DataJpaTest
public class ArticleRepositoryTests {

    @Autowired
    UserRepository userRepository;

    @Test
    public void saveAUser_thenFindIt() {
        //保存一条用户数据，然后进行查询
        User leili = new
User().setLogin("leili").setFirstName("Lei").setLastName("Li");
        userRepository.save(leili);
        User found = userRepository.findById(leili.getId()).orElse(null);
        assertThat(found).isNotEqualTo(null);
```

```java
        assertThat(leili).isEqualTo(found);
    }

    @Test
    public void saveUserList_thenCountThem() {
        //保存一个用户数据集合，然后查询记录的数据量
        User leili = new User().setLogin("leili").setFirstName("Lei").setLastName("Li");
        User meimeihan = new User().setLogin("meimeihan").setFirstName("Meimei")
                .setLastName("Han");
        User taolin = new User().setLogin("taolin").setFirstName("Tao").setLastName("Lin");
        User jimgreen = new User().setLogin("jimgreen").setFirstName("Jim").setLastName("Green");
        List<User> toSave = Arrays.asList(leili, meimeihan, taolin, jimgreen);
        userRepository.saveAll(toSave);
        Long countThem = userRepository.count();
        assertThat(countThem).isEqualTo(4);
    }

    @Test
    public void saveAUser_thenDeleteIt() {
        //保存一条用户数据，然后删除它
        User leili = new User().setLogin("leili").setFirstName("Lei").setLastName("Li");
        userRepository.save(leili);
        userRepository.delete(leili);
        User found = userRepository.findById(leili.getId()).orElse(null);
        assertThat(found).isEqualTo(null);
    }

    @Test
    public void saveUserList_thenDeleteThem() {
        //保存一个用户数据集合，然后删除所有记录
        User leili = new User().setLogin("leili").setFirstName("Lei").setLastName("Li");
        User meimeihan = new User().setLogin("meimeihan").setFirstName("Meimei").setLastName("Han");
        User taolin = new User().setLogin("taolin").setFirstName("Tao").setLastName("Lin");
        User jimgreen = new User().setLogin("jimgreen").setFirstName("Jim").setLastName("Green");
        List<User> toSave = Arrays.asList(leili, meimeihan, taolin, jimgreen);
        userRepository.saveAll(toSave);
        userRepository.deleteAll();
        Long countThem = userRepository.count();
        assertThat(countThem).isEqualTo(0);
    }
}
```

6.4.3 @SpringBootTest 以及其他一些注解

@SpringBootTest 用于加载整个容器的上下文。该注解默认情况下并不会将服务真实地启动，如果需要改变这一行为，需要在使用之处修改 webEnvironment 属性。webEnvironment 属性的可选项如下：

- MOCK（默认）：加载 ApplicationContext 并提供模拟的网络环境。使用该项将不会启动服务。
- RANDOM_PORT：加载 WebServerApplicationContext 并提供真实的网络环境。服务将监听一个随机端口。
- DEFINED_PORT：加载 WebServerApplicationContext 并提供真实的网络环境。服务将监听一个已声明的端口。
- NONE：加载 ApplicationContext 但不提供任何网络环境。

其他自动配置注解：

- @JdbcTest：用以测试基于 JDBC 的组件。
- @DataMongoTest：用于测试 MongoDB 应用。它将配置一个内存嵌入式 MongoDB，配置 MongoTemplate、Spring Data MongoDB 的各 Repository 以及@Document 类。
- @DataRedisTest：用于测试 Redis 应用。它将扫描@RedisHash 类并配置 Spring Data Redis 的各 Repository。

第 7 章

部署与运维

在对程序编写了完备的测试用例之后，软件开发的生命周期到达了最终阶段——部署与运维。随着开发以及项目交付的模式不断更迭，部署以及运维的重要性渐渐被拔高。现代的开发模式强调高效与敏捷，方便快捷的部署并且稳定地运行在其中是不可或缺的一环。

本章将介绍 Spring Boot 程序的部署以及相关工具的使用，涉及的知识点如下：

- Spring Boot 程序的打包、发布与部署
- 程序健康状态监控
- 开发者工具的使用

7.1 发布与部署

为了适应不同的运行环境，Spring Boot 程序的发布方式也很多样。本节将围绕程序发布这一点展开介绍，主要涉及发布的方式、流程、配置以及一些发布相关的工具。

7.1.1 Spring Boot 自身的打包方式 1——可执行 jar 文件

Spring Boot 自身支持的打包方式有两种：可执行 jar 格式和可部署于传统 Web 容器的 war 格式。本小节介绍第 1 种。

jar 文件是一种后缀名为 jar 的打包文件格式，打包内容包含工程已编译的依赖库、资源文件、元数据文件以及压缩好的应用程序类文件。一个 Spring Boot 工程的 jar 文件结构可能是这样的：

```
example.jar
 |
 +-META-INF
```

```
        |   +-MANIFEST.MF
        +-org
        |   +-springframework
        |       +-boot
        |           +-loader
        |               +-<spring boot loader classes>
        +-BOOT-INF
            +-classes
            |   +-mycompany
            |       +-project
            |           +-YourClasses.class
            +-lib
                +-dependency1.jar
                +-dependency2.jar
```

应用程序类将会放在 BOOT-INF/classes 目录中，而依赖则会以 jar 包形式放在 BOOT-INF/lib 中。将项目打包为可执行 jar 文件的方法，以基于 Maven 构建为例，使用 spring-boot-maven-plugin 这个 maven 插件即可实现。在 pom.xml 中插入以下内容，之后通过命令 mvn package 即可完成对工程的打包：

```
<build>
    <plugins>
        <plugin>
            <groupId>org.springframework.boot</groupId>
            <artifactId>spring-boot-maven-plugin</artifactId>
        </plugin>
    </plugins>
</build>
```

可执行 jar 文件的部署是最为方便的方式，部署环境只需要满足安装有对应版本的 JDK/JRE 即可。将 jar 文件放置于部署目录，在其目录执行以下命令即可启动对应的 Spring Boot 程序：

```
java -jar Demo.jar
```

7.1.2　Spring Boot 自身的打包方式 2——部署于传统 Web 容器的 war 格式

war 文件与 jar 文件类似，同样是一种归档文件。区别在于 jar 可以是可单独执行的 jar 文件，也可以是类库文件，而 war 文件则仅仅是依赖 Servlet 容器执行的文件。另外，war 文件的目录也略有不同，如下所示：

```
example.war
 |
 +-META-INF
 |   +-MANIFEST.MF
 +-org
 |   +-springframework
 |       +-boot
 |           +-loader
 |               +-<spring boot loader classes>
 +-WEB-INF
```

```
    +-classes
    |  +-com
    |     +-mycompany
    |        +-project
    |           +-YourClasses.class
    +-lib
    |  +-dependency1.jar
    |  +-dependency2.jar
    +-lib-provided
       +-servlet-api.jar
       +-dependency3.jar
```

依赖项放置于 WEB-INF/lib 中，在运行可单独执行的文件时需要部署于 Servlet 容器中，而不需要的依赖被放置于 WEB-INF/lib-provided 中。

因为 war 文件这一打包方式并不是默认选项，所以需要在默认形式的基础上做一点改动。首先，工程的主类需要继承 SpringBootServletInitializer 并覆盖其 configure 方法，示例如下：

```java
@SpringBootApplication
public class DemoApplication extends SpringBootServletInitializer {

    @Override
    protected SpringApplicationBuilder configure(SpringApplicationBuilder builder) {
        return builder.sources(DemoApplication.class);
    }

    public static void main(String[] args) {
        SpringApplication.run(DemoApplication.class, args);
    }

}
```

下一步需要在 pom.xml 中声明打包方法为 war：

```xml
<packaging>war</packaging>
```

因为 Spring Boot 自带嵌入式 Servlet 容器的缘故，需要确保 Servlet 容器不受 war 文件干扰，需要将嵌入式 Servlet 容器（例如 Tomcat、Jetty、Undertow）的依赖标记为已提供（provided）。以 Maven 为例：

```xml
<dependency>
    <groupId>org.springframework.boot</groupId>
    <artifactId>spring-boot-starter-tomcat</artifactId>
    <scope>provided</scope>
</dependency>
```

最后，与 jar 格式的打包相同，添加 spring-boot-maven-plugin 插件并使用 mvn package 命令进行打包，即可获取工程的 war 文件。

war 格式文件部署依赖于 Servlet 容器，在部署项目前需要保证容器已安装。以 Tomcat 为例，访问其官网 https://tomcat.apache.org/，选择任意发行版（例如 9.0.39Released）下载并安装。安装完成后文件目录如下：

```
├─bin
```

```
├─conf
├─lib
├─logs
├─temp
├─webapps
├─work
│   ...
```

执行 bin 文件夹中的启动脚本 startup.sh/startup.bat 即可启动 Tomcat 容器。之后将打包好的 war 文件放置于 webapps 文件夹内，Tomcat 将会自动解压 war 文件，并将其发布为 Web 服务。例如，demo.war 为打包后的 war 文件，Tomcat 将会将其内容解压至 webapps/demo，对应的服务也将映射到 http://{tomcat-hostname}:{tomcat-portname}/demo 这一路径下。

7.1.3 更现代的发布流程 1——基于 Docker 的发布与部署

上两个小节的流程非常简单，也足以满足需求。但对于一套力求严谨的软件交付流程而言，这些方式显得不够"现代"。接下来将介绍两种现代的发布部署流程：基于 Docker 的发布与部署以及基于 RPM 的发布与部署。本小节先介绍第一种。

Docker 是一个平台即服务的产品（PaaS），使用操作系统级虚拟的容器化技术帮助用户实现软件服务以互相隔离的形式高效运行。在立刻使用之前需要知道三个基本概念：Dockerfile、镜像以及容器。可以简单地把镜像理解为一个可执行程序，容器是运行中的进程。开发人员在完成软件的开发之后，需要将原工程打包成镜像，在这个过程中需要依赖 Dockerfile，它起到一个类似于清单的作用。

基于 Docker 进行发布与部署之前需要完成 Docker 环境的部署，具体流程因不同操作系统而存在差异。

完成 Docker 环境部署后，可以着手开始进行发布与部署。流程如下：

（1）按照可执行 jar 文件的发布流程操作，将项目以 jar 形式打包。
（2）于项目工程的根目录编写 Dockerfile，示例如下：

```
FROM openjdk:8-jdk-alpine
RUN addgroup -S spring && adduser -S spring -G spring
USER spring:spring
ARG JAR_FILE=target/*.jar
COPY ${JAR_FILE} app.jar
ENTRYPOINT ["java","-jar","/app.jar"]
```

Dockerfile 的内容主要用于描述构建一份 Docker 镜像所需要做的配置，下面对以上内容做出简单的解释：

- FROM：该定制镜像是基于 openjdk:8-jdk-alpine 的，后续的工作都是在该基础镜像上完成以得出所需的定制镜像。
- RUN：用于执行后续的命令。示例中的命令表示将添加名为 spring 的用户以及用户组，用于规避使用 root 用户运行服务将面对的风险。
- USER：用于指定运行服务的用户。

- ARG：声明构建参数。
- COPY：复制指令，从上下文目录中复制文件或者目录到容器里指定路径。
- ENTRYPOINT：执行指令。使用方式 ENTRYPOINT ["<executable>","<par-am1>", "<param2>",...]，将参数内容发送至目标程序 executable，执行目标程序。

（3）在 Dockerfile 文件的存放目录下，执行构建命令。

```
docker build -t demo:v1 .
```

> **注　意**
> 命令的结尾有一个"."，代表本次执行的上下文路径。

demo 与 v1 分别代表镜像的名称以及镜像的标签，构建过程如图 7.1 所示。

```
D:\development\cron-task>docker build -t demo:v1 .
Sending build context to Docker daemon  32.78MB
Step 1/6 : FROM openjdk:8-jdk-alpine
8-jdk-alpine: Pulling from library/openjdk
e7c96db7181b: Pull complete
f910a506b6cb: Pull complete
c2274a1a0e27: Pull complete
Digest: sha256:94792824df2df33402f201713f932b58cb9de94a0cd524164a0f2283343547b3
Status: Downloaded newer image for openjdk:8-jdk-alpine
 ---> a3562aa0b991
Step 2/6 : RUN addgroup -S spring && adduser -S spring -G spring
 ---> Running in 450698fc20e9
Removing intermediate container 450698fc20e9
 ---> 16ea481859b3
Step 3/6 : USER spring:spring
 ---> Running in 8cc14e026b0b
Removing intermediate container 8cc14e026b0b
 ---> c750433c5636
Step 4/6 : ARG JAR_FILE=target/*.jar
 ---> Running in 2c27dd9a34ee
Removing intermediate container 2c27dd9a34ee
 ---> 4ab3601ba861
Step 5/6 : COPY ${JAR_FILE} app.jar
 ---> c74eaa3207f4
Step 6/6 : ENTRYPOINT ["java","-jar","/app.jar"]
 ---> Running in 82bd862b3b6f
Removing intermediate container 82bd862b3b6f
 ---> b3e4e3f1e368
Successfully built b3e4e3f1e368
Successfully tagged demo:v1
```

图 7.1　Docker 镜像构建过程

（4）将构建好的镜像推送至镜像仓库进行管理。推送前需要保证已完成仓库的登录，登录命令如下：

```
docker login
```

推送命令如下：

```
docker push
```

（5）在需要部署服务的机器上部署好 Docker 环境，通过拉取命令拉取仓库中的服务镜像。拉取命令如下：

```
docker pull
```

（6）使用启动命令以启动镜像对应的容器。启动命令如下：

```
docker run -p 8080:8080 demo/v1
```

至此，基于 Docker 的发布与部署便完成了。像这样的部署方式相较于基础的部署流程，减少了手工维护的程度，这样可以更方便地实现标准化和规范化。

7.1.4　更现代的发布流程 2——基于 RPM 的发布与部署

RPM（Redhat Package Manager）是一个强大的软件包管理工具，用来安装、卸载、校验和更新 Linux 系统上的软件包。因此，该发布方式适用于使用 RPM 进行包管理的 Linux 系统（比如 CentOS）。使用该流程有不少前置依赖，负责编译的环境需要同时装有 JDK、Maven 以及 rpm-build 工具。如果依赖项已安排妥当，便可以开始发布与部署了，具体流程如下：

（1）在 pom.xml 中使用 rpm-maven-plugin 替换 spring-boot-maven-plugin。示例代码如下：

```
<build>
  <plugins>
 ...

   <plugin>
     <groupId> org.codehaus.mojo </ groupId>
     <artifactId> rpm-maven-plugin </ artifactId>
     <version> 2.2.0 </ version>
     <extensions> true </ extensions>
      <configuration>
     ...
     </ configuration>
   </ plugin>
...
  </ plugins>
</ build>
```

另外需要在 pom.xml 中声明打包方式为 rpm：

```
<packaging> rpm </ packaging>
```

（2）下面的改动都集中于 rpm-maven-plugin 的配置项。第一个配置是<group>：

```
<group> $ {project.groupId} </ group>
```

（3）配置服务的安装目录以及相关内容：

```
<mapping>
  <directory>/var/demo-application/</directory>
  <filemode>755</filemode>
  <username>demo-application</username>
  <groupname>demo-application</groupname>
  <dependency>
```

```xml
            <stripVersion>true</stripVersion>
            <includes>
              <include>${project.groupId}:demo-application-app</include>
            </includes>
        </dependency>
</mapping>
```

（4）完成以上内容后，已经满足打包一个 rpm 软件包的条件。但是还留有一些问题：

- 以上声明中包含的用户以及用户组可能在所部署环境下不存在。
- 程序不会以系统服务的形式运行，换句话说不能做到开机自启动等操作。

RPM 软件包在安装过程中允许使用脚本，因此可以针对这些需求另外编写脚本。这些问题可以分别编写两个脚本，分别是安装前执行的脚本 preinstall.sh 和安装后执行的脚本 postinstall.sh。这两份脚本内容如下：

preinstall.sh：

```bash
# 添加"demo-application"用户以及用户组
/usr/sbin/useradd -c "Demo Application" -U \
      -s /sbin/nologin -r -d /var/demo-application demo-application 2> /dev/null || :
```

postinstall.sh：

```bash
# 确保服务已启动
if [ $1 -eq 1 ] ; then
  # 初始安装
  systemctl enable demo-application.service >/dev/null 2>&1 || :
fi
```

在 src/main/resources 目录下编写好这两份脚本后，在 pom.xml 中使用 preinstallScriptlet 以及 postinstallScriptlet 将它们配置好：

```xml
<preinstallScriptlet>
    <scriptFile>src/main/resources/preinstall.sh</scriptFile>
    <fileEncoding>utf-8</fileEncoding>
    <filter>true</filter>
</preinstallScriptlet>
<postinstallScriptlet>
    <scriptFile>src/main/resources/postinstall.sh</scriptFile>
    <fileEncoding>utf-8</fileEncoding>
    <filter>true</filter>
</postinstallScriptlet>
```

（5）将程序注册成服务需要编写对应的 service 文件 demo-application.service，该文件也可以在发布前编写好，待安装过程中再由 RPM 复制至 /etc/systemd/system 内。在 src/main/resources 目录下创建文件夹 systemd，以及在该目录下创建文件 demo-application.service：

```
[Unit]
Description=demo-application
After=syslog.target
[Service]
User=demo-application
```

```
ExecStart=/var/demo-application/demo-application-app.jar
SuccessExitStatus=143
[Install]
WantedBy=multi-user.target
```

在 pom.xml 中新增一个 mapping 项：

```xml
<mapping>
  <directory>/etc/systemd/system</directory>
  <filemode>755</filemode>
  <username>root</username>
  <groupname>root</groupname>
  <sources>
    <source>
      <location>src/main/resources/systemd</location>
    </source>
  </sources>
</mapping>
```

（6）通过 mvn package 命令即可完成打包工作，打包结果为 rpm 格式的软件包文件。通过 yum localinstall 可以实现本地安装，也可以选择将 rpm 推送到 yum 仓库中进行管理。

7.1.5　多环境配置

开发过程中或许会碰到这样一种麻烦，一个程序在软件开发的不同阶段中需要被部署到不同的环境中（比如开发环境、测试环境、演示环境以及生产环境），各个环境之间的配置项存在差异。如果这些配置只依赖于手工修改，不能被很好地管理起来的话，每增加一行配置，维护难度就会上一层台阶，最终会造成无法预料的可怕局面。

在 Spring Boot 开发中，Spring 提供了 Spring Profile 功能。它允许我们将环境这一变量抽离出来，每一种 profile 对应一套配置，指定某 profile 后，部署文件内将仅启用对应的配置文件。这一功能再结合 maven profile 功能，可以完美解决多环境带来的配置问题。

（1）首先针对不同环境创建不同的配置文件 application-${env}.yml。如以下示例所示，当前存在开发环境（dev）和生产环境（prod）。不同环境的服务端口号不同。

application-dev.yml：

```
server:
  port: 8080
```

application-prod.yml：

```
server:
  port: 8081
```

（2）在 application.yml 中使用 spring profile 配置：

```
spring:
  profiles:
    active: @activatedProperties@
```

（3）在 pom.xml 添加 profile 以及对应的 activatedProperties 变量：

```xml
<profiles>
    <profile>
        <id>dev</id>
        <activation>
            <activeByDefault>true</activeByDefault>
        </activation>
        <properties>
            <activatedProperties>dev</activatedProperties>
        </properties>
    </profile>
    <profile>
        <id>prod</id>
        <properties>
            <activatedProperties>prod</activatedProperties>
        </properties>
    </profile>
</profiles>
```

dev 为默认的 profile，它们的 activatedProperties 与 profile 名相同。该参数将会在打包后注入到 application.yml 的 spring.profile.active。

（4）程序打包，打包命令：

```
mvn clean package -P <profile-id>
```

使用 idea 可以通过勾选选项启用对应的 profile，如图 7.2 所示。

图 7.2 启用 profiles 后端 maven 插件

（5）启动程序，可以在启动过程的日志中看到当前所使用的 profile。

7.2 运行监控

运行监控这一环节在软件开发的过程中是容易被忽视的一个环节。大多数开发人员的关注点集中于功能实现以及性能，程序健壮性通常交给各项测试去验证。然而，真实的运行环境终归是复杂多变的，即便概率再小，意外情况总会发生。为了预警风险并且解决问题，让程序拥有运行监控的功能是非常必要的。

7.2.1 使用 Spring Boot Actuator 查看运行指标

Actuator 是由 Spring Boot 提供用于程序运维的相关工具。使用该工具可以很方便地对程序状态进行监控，包括各种指标、流量以及数据库状态等。集成该工具过后，可以通过 HTTP 端点以及 JMX Bean 的方式获取它收集到的信息。

启用 Spring Boot Actuator 只需要添加对应的依赖项即可。Spring Boot Actuator 的依赖项如下：

> **注　意**
>
> 监控的内容涉及程序的安全性，在正式使用过程中，请务必再另外加上 spring-boot-start-security 依赖。

```xml
<dependency>
    <groupId>org.springframework.boot</groupId>
    <artifactId>spring-boot-starter-actuator</artifactId>
</dependency>
```

依赖项添加成功之后，可以通过/actuator 路径查看到当前可用的 Actuator 端点，默认开启的有 health 以及 info。默认内容如下：

```
{
  "_links": {
    "self": {
      "href": "http://localhost:8080/actuator",
      "templated": false
    },
    "health": {
      "href": "http://localhost:8080/actuator/health",
      "templated": false
    },
    "health-path": {
      "href": "http://localhost:8080/actuator/health/{*path}",
      "templated": true
    },
    "info": {
      "href": "http://localhost:8080/actuator/info",
      "templated": false
    }
  }
}
```

Actuator 的可用端点 ID 如下所示：

- auditevents：公开当前应用程序的审核事件信息。
- beans：显示应用程序中所有 Spring Bean 的完整列表。
- caches：公开可用的缓存。

- conditions：显示在配置和自动配置类上评估的条件以及它们匹配或不匹配的原因。
- configprops：显示所有的整理列表@ConfigurationProperties。
- env：公开 Spring 的属性 ConfigurableEnvironment。
- flyway：显示所有已应用的 Flyway 数据库迁移。
- health：显示应用程序运行状况信息。
- httptrace：显示 HTTP 跟踪信息（默认情况下，最近的 100 个 HTTP 请求-响应）。
- info：显示应用程序信息。
- integrationgraph：显示 Spring Integration 图。
- loggers：显示和修改应用程序中记录器的配置。
- liquibase：显示已应用的所有 Liquibase 数据库迁移。
- metrics：显示当前应用程序的"指标"信息。
- mappings：显示所有@RequestMapping 路径的整理列表。
- scheduledtasks：显示应用程序中的计划任务。
- sessions：允许从 Spring Session 支持的会话存储中检索和删除用户会话。
- shutdown：使应用程序正常关闭。
- threaddump：执行线程转储。

通过在 application.yml 中修改相关配置可以启用或者关闭相应端点，配置示例如下：

- management.endpoints.jmx.exposure.exclude：JMX 排除项。
- management.endpoints.jmx.exposure.include：JMX 包含项（默认值*）。
- management.endpoints.web.exposure.exclude：web 端点排除项。
- management.endpoints.web.exposure.include：web 端点包含项（默认 info，health）。

掌握以上内容便可以使用 Actuator 开放端点，以向外提供监控数据查询的途径。

7.2.2　集成 Prometheus

仅仅提供数据查询的接口，并不便于完成对程序的监控。下面将通过整合 Prometheus，完善对程序的监控。Prometheus 是一款开源的系统监控和警报工具，它有以下特点：

- 一个多维数据模型，它包含由指标名以及键-值对（Key Value Pair）区分的时序数据。
- PromQL，一种灵活的查询语言。
- 不依赖分布式存储，单服务节点是自洽的。
- 通过 HTTP 拉取时序数据。
- 通过网关支持时序数据的推送。
- 通过服务发现或静态配置发现监控目标。
- 多种图形和仪表盘支持模式。

使用 Prometheus 可以实现对 Actuator 提供的指标数据的收集、搜索以及展示。该工具是以独立服务的形式运行，因此需要单独部署。推荐使用 Docker 进行部署，步骤如下：

（1）拉取 Prometheus 镜像：

```
docker pull prom/prometheus
```

（2）编写 Prometheus 的配置文件 prometheus.yml：

```
#全局配置
global:
  #拉取间隔，默认1分钟
  scrape_interval:
  external_labels:
monitor: 'demo-monitor'
#拉取配置
scrape_configs:
  #任务名
  - job_name: 'prometheus'
    scrape_interval: 5s
static_configs:
  #采集地址
    - targets: ['localhost:9090']
  - job_name: 'demo'
    #采集指标的路径
    metrics_path: '/actuator/prometheus'
    scrape_interval: 5s
    static_configs:
      - targets: ['localhost:8080']
```

（3）启动容器：

```
docker run --rm -p 9090:9090 /
-v c:/prometheus/prometheus.yml:/etc/prometheus/prometheus.yml /
prom/prometheus
```

参数说明：

- --rm：表示将清理历史运行数据。
- -p：用于指定端口映射。
- -v：用于文件挂载，此处将 Windows 系统下的 C:/prometheus/prometheus.yml 挂载到容器中的 /etc/prometheus/prometheus.yml。

（4）完成指令的输入之后，可以通过 http://localhost:9090 访问 Prometheus 的管理界面以确认服务的运行情况。在表达式输入框内输入指标名或者构造 PromQL 后，单击 Execute 即可获取对应数值以及简易的图表信息，如图 7.3 所示。

（5）为 Spring Boot 启用 prometheus 端口。引入依赖如下：

```
<dependency>
  <groupId>io.micrometer</groupId>
  <artifactId>micrometer-registry-prometheus</artifactId>
</dependency>
```

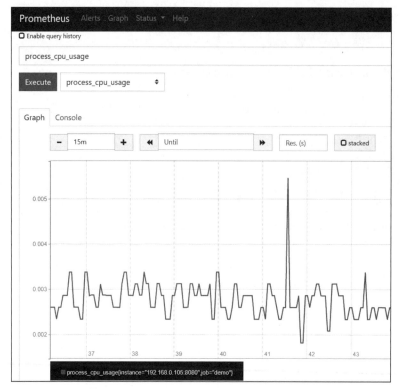

图 7.3　Prometheus 控制台界面

在 application.properties 中加入配置启用端点：

```
management.endpoints.web.exposure.include=*
# 或者
management.endpoints.web.exposure.include=prometheus
```

启动过后可以通过路径 http://localhost:8080/actuator/prometheus 查看端口是否成功开启。

（6）确认 Prometheus 是否可以成功拉取指标数据。通过 Status→Targets 可以查看所监控服务的运行情况，如图 7.4 和图 7.5 所示。

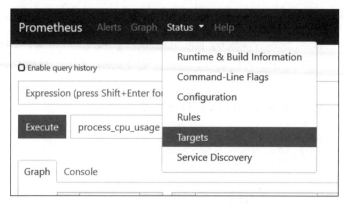

图 7.4　查看所监控服务的运行情况

图 7.5　监控服务健康情况

7.2.3　Grafana 实现可视化监控

Prometheus 为开发人员提供了指标的记录、搜索以及展示，已经具备了基础的监控功能。不过并不是所有人对于 PromQL 以及各项指标都了如指掌，Prometheus 所提供的图表功能也略显单薄。Grafana 可以填补 Prometheus 这一空隙，它是一款开源的度量分析与可视化套件，通过访问数据源以展示各种自定义报表。它的 UI 十分灵活，有丰富的插件和模板。它功能强大，常用在时序数据的监控方面。

与 Prometheus 类似，Grafana 也是一个需要独立部署的服务。同样推荐使用 Docker 进行部署，步骤如下：

（1）拉取镜像，命令如下：

```
docker pull grafana/grafana
```

（2）创建用于存储 Grafana 数据的文件夹 grafana。

（3）在启动时将 Grafana 挂载上去，命令如下：

```
docker run -rm -p 3000:3000 --name=grafana -v c:/grafana:/var/lib/grafana grafana/grafana
```

（4）访问 Grafana，路径为 http://localhost:3000，如图 7.6 所示。

图 7.6　Grafana 登录页面

默认登录账号和密码为 admin/admin，第一次登录将会弹出重置密码的提示，如图 7.7 所示。

图 7.7　重置密码提示

（5）完成密码的重置后将跳转至主页，开始设置数据源。通过单击左侧"Configuration"中的"Data Sources"添加数据源，如图 7.8 所示。

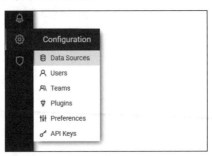

图 7.8　配置数据源

在 HTTP 栏填入 URL 并选择 Access 为 Browser 后单击 Save&Test 按钮，如图 7.9 所示。

图 7.9　Grafana 数据源配置页面

（6）配置仪表盘。Grafana 拥有众多的社区为开发人员提供帮助，仪表盘可以使用社区里的模板配置。通过单击左侧菜单"Create"中的"Import"，进入导入配置页面，如图 7.10 所示。

图 7.10　导入 Dashboard 配置

仪表盘模板可以通过 https://grafana.com/grafana/dashboards 页面查到。可以使用 Spring Boot 为关键字检索推荐的模板，如图 7.11 所示。

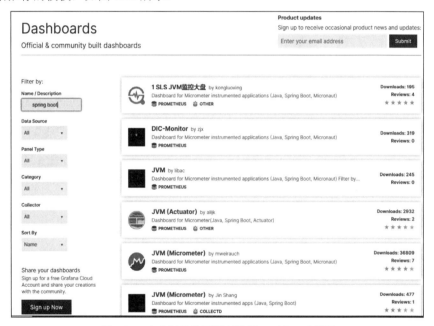

图 7.11　官方以及社区所创作的 Dashboard 模板

将对应 ID 或 URL 填入配置页的输入框，选择 Prometheus 作为数据源即可完成仪表盘模板的导入。

（7）完成导入后，便可以使用仪表盘查看程序的各项运行指标，如图 7.12 所示。

至此，基于 Spring Boot Actuator、Prometheus 以及 Grafana 的可视化监控功能搭建完成。

图 7.12　社区提供的 SLS 仪表盘

7.3　Spring Boot 开发者工具

Spring Boot 给开发人员带来了快速、便捷的开发体验。为了让这一体验能更进一步，Spring 还发布了开发者工具 spring-boot-devtools。该工具解决了开发过程中将会遇到的一些烦琐问题，让开发人员能更好地关注程序开发本身。

7.3.1　整合 spring-boot-devtools

在项目中整合 spring-boot-devtools 与添加其他 Spring Boot 模块一样简单。加入以下依赖即可：

```xml
<dependencies>
    <dependency>
        <groupId>org.springframework.boot</groupId>
        <artifactId>spring-boot-devtools</artifactId>
        <optional>true</optional>
    </dependency>
</dependencies>
```

> **注　意**
>
> spring-boot-devtools 所带来的优化均针对开发以及测试环境。在这些环境上运行的、所需要具备的条件与在生产环境上存在差异。因此，在为生产环境提供软件包时，务必将该依赖排除在外。Spring Boot 团队对此做了一些工作，比如在运行"完全打包"的应用程序时，spring-boot-devtools 将会被禁用。"完全打包"的应用程序指的是使用 java -jar 命令启动的，或者使用特殊的类加载器启动的程序。在依赖中设置"optional"为 true 是 Spring Boot 团队认定的"最佳实践"。除此之外，也可以借助 maven profile 功能，将开发环境的依赖项与生产环境的依赖项区分开来。

7.3.2 自动配置

Spring Boot 为了提高程序在生产环境下的性能，内置了诸多配置，比如默认情况下会启用缓存以提高性能。像是模板引擎中就用到了这样的优化方式，但是在开发环境中，尽快地查看到修改结果相对而言更为重要。通常会通过在 application.properties 中添加配置 spring.thymeleaf.cache=false 为 thymeleaf 禁用默认的缓存。在引入开发者工具 spring-boot-devtools 之后，这些操作将由工具帮助开发人员自动完成。

7.3.3 热部署

在日常开发过程中一定会碰到修改程序并重启的情况。如果依赖项过多导致加载依赖项非常耗时，那么频繁的重启会浪费掉很多时间。spring-boot-devtools 为了解决这个问题带来了"热部署"功能，该功能可以实现应用程序的快速重启。

spring-boot-devtools 将会为应用创建两个类加载器，分别是加载不变类（例如来自第三方 jar 的类）的基础类加载器和加载可变类（当前正在开发的类）的重启类加载器。每当重启时，将丢弃重启类加载器，并创建一个新的类加载器。这种方式相较于应用平时使用的"冷启动"要快得多，毕竟基础类加载器已经处于可用的状态。

spring-boot-devtools 的热部署功能将监听当前项目的 classpath，如果 classpath 上有文件更改，热部署将被激活，程序随之自动重启。要流畅使用该功能还需要对 IDE 做一些配置使 IDE 可用，在修改过后进行自动编译。以 IDEA 为例，流程如下：

（1）开启自动编译项目，路径为"File"→"Settings"→"Build,Ex-ecution,Deployment"→"compiler"，如图 7.13 所示。

图 7.13　开启自动构建项目

（2）使用快捷键"Shift+Ctrl+A"激活 Find Action 搜索框。

（3）在搜索框中输入 Registry，并单击搜索到的 Registry 选项，如图 7.14 所示。

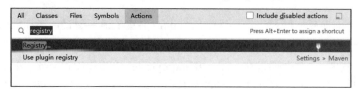

图 7.14　搜索 Registry

（4）在 Registry 弹出框中搜索属性"compiler.automake.allow.when.app.running"，并勾选，如图 7.15 所示。

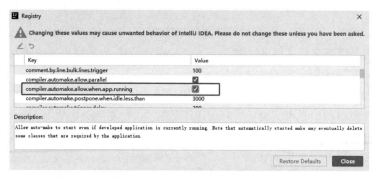

图 7.15　勾选 compiler.automake.allow.when.app.running

事实上，并不总是需要在文件更改之后重新启动程序，比如一些静态文件。例如模板、CSS、二进制文件这样的静态文件可以通过 LiveReload 功能实现快速更改。因此可以在配置中，将这部分文件所在路径从重启的监听路径中排除。配置示例如下：

```
spring.devtools.restart.exclude=static/**,public/**
```

可以将特定路径排除在监听的范围，也可以将另外的特定路径加入到监听的范围内。将附加的路径加入到监听范围内的配置项为：

```
spring.devtools.restart.additional-paths
```

7.3.4　LiveReload 插件支持静态资源的及时更新

spring-boot-devtools 内包含一个嵌入式 LiveReload 服务器，该服务器用于静态资源更改后的更新。要使用该功能，需要另外在浏览器中获取支持该功能的 LiveReload 插件。

如果不需要启用 LiveReload 服务器，可以在 application.properties 中加入以下配置：

```
spring.devtools.livereload.enabled=false
```

7.3.5　全局配置

spring-boot-devtools 支持在计算机中设置一个全局配置。该配置的内容将应用于当前计算机所有启用了 devtools 的 Spring Boot 应用程序。配置路径为$HOME 文件夹下的.spring-boot-devtools.properties 文件。Windows 用户使用该功能，或许需要另外设置环境变量 HOME。

7.3.6 远程应用

spring-boot-devtools 提供了开箱即用的远程调试功能，要使用该功能，需要将 spring-boot-devtools 打包为程序的一部分。可以通过 maven 插件中的 excludeDevtools 配置来实现：

```xml
<build>
    <plugins>
        <plugin>
            <groupId>org.springframework.boot</groupId>
            <artifactId>spring-boot-maven-plugin</artifactId>
            <configuration>
                <excludeDevtools>false</excludeDevtools>
            </configuration>
        </plugin>
    </plugins>
</build>
```

为了使通过 HTTP 的远程调用生效，需要执行以下步骤：

（1）在启动服务端程序的时候添加以下参数：

```
-Xdebug -Xrunjdwp:server=y,transport=dt_socket,suspend=n
```

（2）在本地 IDE 中，打开启动配置，填入以下内容：

- 主类：org.springframework.boot.devtools.RemoteSpringApplication
- 程序参数：应用程序的 URL，例如 http://localhost:8080

如图 7.16 所示。

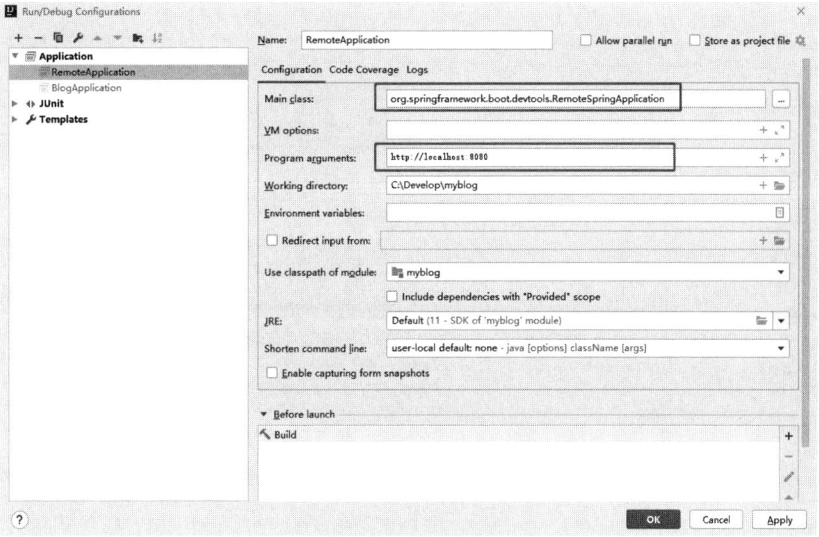

图 7.16 配置远程应用

（3）启动主类为 RemoteSpringApplication 的启动项，以启用远程应用客户端。

完成远程应用客户端的启动过后，可以实现远程自动更新（与本地热部署类似）。本地项目的改动将会由远程应用客户端推送给服务端程序，并自动完成更新。另外还可以通过该启动项实现远程调试，步骤与本地调试类似。

远程调试的默认端口为 8000，可以使用以下配置对其进行更改：

```
spring.devtools.remote.debug.local-port=8010
```

另外，为保证程序的安全性，还可以使用以下配置为远程应用设置秘钥：

```
spring.devtools.remote.secret=mysecrt
```

第 8 章

实战 1：基于 STOMP 协议的聊天服务

本章将介绍一个基于 STOMP 协议的聊天服务端程序。该聊天服务端程序的整体实现基于 Spring Boot 以及 Spring 相关的项目，实现了一个聊天服务最基础的功能，包括注册登录、群聊、聊天记录等。

本章内容包含程序架构设计、主要模块的实现以及程序测试，可以帮助读者更好地了解聊天程序主要的开发流程。涉及的知识点如下：

- 基于 Spring Boot 的项目构建
- WebSocket 以及 STOMP 协议
- WebSocket 整合 Spring Security

8.1 架构设计

基于 STOMP 协议的聊天服务，整体分为三大模块，分别为消息模块、存储模块以及安全模块。消息模块基于 STOMP 协议实现，安全模块基于 Spring Security 实现，存储模块基于 MongoDB 实现。程序的架构图如图 8.1 所示。

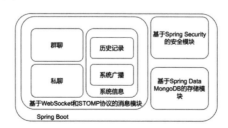

图 8.1 聊天服务架构图

聊天服务主要基于 STOMP 协议，另外辅以 HTTP 协议与客户端进行交互。存储模块主要用于用户信息以及聊天信息的存储。安全模块对需要进行保护的 HTTP 端点采取相应措施之外，还将会对 WebSocket 端点提供防护。

8.2　框架搭建

在程序开发开始之前，需要进行整体框架的搭建。根据程序的整体架构判断，所需的依赖主要包括以下的项目：

- spring-boot-starter-websocket
- spring-boot-starter-security
- spring-boot-starter-data-mongodb
- spring-security-messaging

打开 https://start.spring.io/ 选择所需的依赖项，如图 8.2 所示。

图 8.2　Spring Initializr

单击左下角 "GENERATE" 按钮生成项目压缩包并自动完成下载。并不是所有依赖都包含在 Spring Initializr 中，解压压缩包后需要简单确认一下 pom.xml 的内容，特别是依赖部分。示例如下：

```
<dependencies>
    <dependency>
        <groupId>org.springframework.boot</groupId>
        <artifactId>spring-boot-starter-web</artifactId>
    </dependency>
```

```xml
<!--lombok 插件支持，可选项-->
<dependency>
    <groupId>org.projectlombok</groupId>
    <artifactId>lombok</artifactId>
    <optional>true</optional>
</dependency>
<dependency>
    <groupId>org.springframework.boot</groupId>
    <artifactId>spring-boot-starter-test</artifactId>
    <scope>test</scope>
    <exclusions>
        <exclusion>
            <groupId>org.junit.vintage</groupId>
            <artifactId>junit-vintage-engine</artifactId>
        </exclusion>
    </exclusions>
</dependency>
<!--WebSocket 依赖-->
<dependency>
    <groupId>org.springframework.boot</groupId>
    <artifactId>spring-boot-starter-websocket</artifactId>
</dependency>
<!--安全框架-->
<dependency>
    <groupId>org.springframework.boot</groupId>
    <artifactId>spring-boot-starter-security</artifactId>
</dependency>
<!--Mongodb 存储-->
<dependency>
    <groupId>org.springframework.boot</groupId>
    <artifactId>spring-boot-starter-data-mongodb</artifactId>
</dependency>
<dependency>
    <groupId>org.springframework.boot</groupId>
    <artifactId>spring-boot-autoconfigure</artifactId>
</dependency>
<!--WebSocket 集成 Spring Security 相关-->
<dependency>
    <groupId>org.springframework.security</groupId>
    <artifactId>spring-security-messaging</artifactId>
</dependency>
</dependencies>
```

完成项目的创建后，下一步将对项目的代码结构做基础的规划。以包为单位进行划分，内容如下：

- config：包含配置相关类。
- controller：包含控制器相关类，其中包括基于 HTTP 的 REST 控制器以及基于 STOMP 的控制器。
- entity：包含实体类。
- model：包含模型类。
- repository：包含 Repository 类。

- security：包含安全相关的类。

预先规划好清晰的代码结构，有利于后面开发工作的展开。

8.3 功能实现

消息模块作为聊天程序的核心部分，首先需要实现。基于可靠性、生态支持以及开发难度等角度考虑，该消息模块基于 STOMP 协议进行开发。主要用于实现群聊以及系统消息的功能。在完成消息模块的实现之后，再围绕该模块进行附加功能的扩展。

8.3.1 了解 WebSocket 协议

WebSocket 协议是一种通信协议。与 HTTP 有一些类似，它们同属于 OSI 模型的应用层，并依赖于传输层的 TCP 协议。不同的是，HTTP 提供的是单向的通信信道，而 WebSocket 协议提供的是双向的通信信道。OSI 模型如图 8.3 所示。

图 8.3 OSI 模型

HTTP 与 WebSocket 连接对比如图 8.4 所示。

WebSocket 的协议头为 ws://或者 wss://。它既是一个双向通信协议，同时还是一个有状态的协议。这意味着客户端与服务端之间的连接需要保持活动状态，直到被其中一方终止。客户端或服务端中的任意一个关闭连接后，连接将从两端终止。

WebSocket 与 HTTP 协议存在依赖关系。该依赖关系指的是 WebSocket 的请求与握手过程是基于 HTTP 协议的。在实际的实现中，WebSocket 连接是由一个 HTTP 请求"升级"而来的。

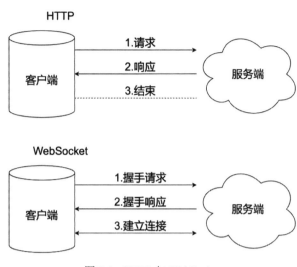

图 8.4　HTTP 与 WebSocket

8.3.2　HTTP 请求升级至 WebSocket 的过程

首先，客户端将发送一个握手请求：

```
GET http://localhost:8080/ws/510/1alwc3mc/websocket HTTP/1.1
Host: localhost:8080
Connection: Upgrade
Upgrade: websocket
Pragma: no-cache
Cache-Control: no-cache
User-Agent: Mozilla/5.0 (Windows NT 10.0; Win64; x64) AppleWebKit/537.36 (KHTML, like Gecko) Chrome/86.0.4240.111 Safari/537.36
Origin: http://localhost:8080
Cookie: JSESSIONID=0DE9FBC170FAB3AFD8FD9D9C55B12F62
Sec-WebSocket-Version: 13
Sec-WebSocket-Key: pWLdwaA2fphhftrEWoUJHg==
Sec-WebSocket-Extensions: permessage-deflate; client_max_window_bits
```

该请求与普通的 HTTP 请求不同之处在于：

```
Connection: Upgrade
Upgrade: websocket
```

这两处将告知服务端，本请求是用于发起 WebSocket 的。

还有 Sec-WebSocket-Key 等属性：

```
Sec-WebSocket-Version: 13
Sec-WebSocket-Key: pWLdwaA2fphhftrEWoUJHg==
Sec-WebSocket-Extensions: permessage-deflate; client_max_window_bits
```

Sec-WebSocket-Version 用于告知服务端所使用的 WebSocket 版本。

Sec-WebSocket-Key 发送给服务端并计算出 Sec-WebSocket-Accept，以帮助客户端确认服务端身份。

Sec-WebSocket-Extensions 用于请求扩展。

服务端响应如下：

```
HTTP/1.1 101
Vary: Origin
Vary: Access-Control-Request-Method
Vary: Access-Control-Request-Headers
Upgrade: websocket
Connection: upgrade
Sec-WebSocket-Accept: H/pGXTc8aiidD16GhlRWzC6SnaQ=
Sec-WebSocket-Extensions: permessage-deflate;client_max_window_bits=15
```

服务端同样返回以下内容，告知客户端请求升级成功：

```
Upgrade: websocket
Connection: upgrade
```

8.3.3 了解 WebSocket 应用场景

WebSocket 相当于 Web 环境下对 HTTP 的补充，使得 Web 应用程序的开发不局限于各项管理系统、博客、门户网站等形式，可以进入到更广阔的领域。常见的应用场景：

- 实时数据展示：一些网站存在对实时数据展示的需求，例如天气数据、比特币交易或者股票信息等。
- 游戏：WebSocket 的出现使得 Web 端游戏程序的开发变得更加方便。
- 实时通信：聊天程序通过 WebSocket 建立一次连接，之后便可以依赖各客户端与服务端之间的连接实现消息的单播或者广播。

8.3.4 集成 WebSocket

JSR 356（Java Specification Requests，Java 规范提案）指定了 Java 开发人员可用于开发 WebSocket 的 API。依赖如下：

```
<dependency>
    <groupId>javax.websocket</groupId>
    <artifactId>javax.websocket-api</artifactId>
    <version>1.1</version>
</dependency>
```

WebSocket 的开发需要对端点进行配置，并监听各端点上的不同事件。实现方式有两种，一种是通过继承 javax.websocket.Endpoint 来实现，另一种通过 API 所提供的注解来实现。注释如下：

- @ServerEndpoint：服务端使用该注解启用对端点的监听。
- @ClientEndpoint：客户端使用该注解。
- @OnOpen：当启动新的 WebSocket 连接时，容器将调用带有@OnOpen 的方法。
- @OnMessage：当消息发送到端点时，容器将调用带有@OnMessage 的方法。
- @OnError：通信出现问题时，将调用带有@OnError 的方法。

- @OnClose：连接关闭时由容器调用的方法。

一个服务端实现的示例代码如下：

```java
@ServerEndpoint(value = "/chat/{username}")
public class ChatEndpoint {

    /*会话*/
    private Session session;
    /*用于维护已创建的端点实例*/
    private static Set<ChatEndpoint> chatEndpoints = new CopyOnWriteArraySet<>();
    /*用于维护在线用户列表*/
    private static HashMap<String, String> users = new HashMap<>();

    @OnOpen
    public void onOpen(Session session) throws IOException {
        //处理连接事件
    }

    @OnMessage
    public void onMessage(Session session, ChatMessage message) throws IOException {
        //处理收信事件
    }

    @OnClose
    public void onClose(Session session) throws IOException {
        //处理断连事件
    }

    @OnError
    public void onError(Session session, Throwable throwable) {
        //处理报错事件
    }

}
```

广播方法的实现：

```java
private static void broadcast(ChatMessage message)
        throws IOException, EncodeException {
    //消息广播
    chatEndpoints.forEach(endpoint -> {
        synchronized (endpoint) {
            try {
                endpoint.session.getBasicRemote().sendObject(message);
            } catch (IOException | EncodeException e) {
                e.printStackTrace();
            }
        }
    });
}
```

在该方法中，将对该类所维护的端点 Set 进行迭代消息发送，以实现广播的效果。

onOpen 方法的实现：

```
@OnOpen
public void onOpen(Session session, @PathParam("username") String username)
        throws IOException, EncodeException {
    //处理连接建立的事件
    this.session = session;
    chatEndpoints.add(this);
    users.put(session.getId(), username);
    //广播用户上线信息
    ChatMessage message = new ChatMessage();
    message.setSender(username);
    message.setContent("Connected!");
    broadcast(message);
}
```

在 onOpen 方法中将可以获取到一个 Session 对象，用于管理服务端到客户端之间的会话。需要在该方法内将其保存下来，在广播或者单播的时候将会用到它。

onMessage 方法的实现：

```
@OnMessage
public void onMessage(Session session, ChatMessage message)
        throws IOException, EncodeException {
    //收到客户端发送的消息后，将信息进行广播
    message.setSender(users.get(session.getId()));
    broadcast(message);
}
```

onClose 方法的实现：

```
@OnClose
public void onClose(Session session) throws IOException, EncodeException {
    //移除离线信息
    chatEndpoints.remove(this);
    users.put(session.getId(), null);
    //广播离线信息
    ChatMessage message = new ChatMessage();
    message.setSender(users.get(session.getId()));
    message.setContent("Disconnected!");
    broadcast(message);
}
```

8.3.5 使用 STOMP 协议实现消息模块

从上一节的示例中可以看到，基于 WebSocket 的开发主要关注点在于 WebSocket 的各种事件，开发人员针对不同事件编写相应代码。WebSocket 虽然足够简单明了，但这也迫使开发人员在实际开发过程中需要对其进行进一步的封装，这样才能让开发的过程足够高效和顺畅。封装的部分不仅仅包括服务端，同样还涉及客户端。为了尽可能减少开发量，一个通用并且实用的子协议是当前迫切需要的，这就是选用 STOMP 协议的理由。

STOMP 协议是一种基于帧的协议，可以基于 WebSocket 协议进行传输。一帧由命令、可选的标头和可选的消息体组成。STOMP 协议基于文本，也允许进行二进制传输。默认编码为 UTF-8，

也支持使用其他编码替代。STOMP 帧示例如下：

```
COMMAND
header1:value1
header2:value2

Body^@
```

STOMP 协议的命令如下：

- SEND：发送帧。用于将消息发送至目的地。
- SUBSCRIBE：订阅帧。用于注册以收听目的地的消息。
- UNSUBSCRIBE：取消订阅帧。用于删除从目的地的订阅。
- BEGIN：开始帧。用于启用事务。
- COMMIT：提交帧。用于提交当前正在进行的事务。
- ABORT：中止帧。用于回滚正在进行的事务。
- ACK：确认帧。用于确认来自订阅方的消息。
- NACK：非确认帧。用于告知服务端当前客户端未使用该订阅消息。
- DISCONNECT：关闭帧。客户端可以随时关闭与服务端的连接，但不能保证服务端已接收到先前所发送的帧。借用关闭帧可以实现对这方面的确认。

1. Spring Boot 启用 STOMP

启用 STOMP 所需依赖在 8.2 节已介绍，此处不再赘述。

创建/config/WebSocketConfig.java 用于开启 STOMP：

```
@Configuration
@EnableWebSocketMessageBroker
public class WebSocketConfig implements WebSocketMessageBrokerConfigurer {

    @Override
    public void registerStompEndpoints(StompEndpointRegistry registry) {
        registry.addEndpoint("/ws").withSockJS();
    }

    @Override
    public void configureMessageBroker(MessageBrokerRegistry registry) {
        registry.setApplicationDestinationPrefixes("/app");
        registry.enableSimpleBroker("/topic");
    }
}
```

使用以上配置，将"/ws"注册为一个 STOMP 端点，并且让其支持 SockJS（用于在不支持 WebSocket 的浏览器上模拟 WebSocket）。另外，将注册"/app"与"/topic"这两个"目的地前缀"。目的地前缀与@MessageMapping、@SubscribeMapping 注解联合使用，可以控制消息的分发与处理，如图 8.5 所示。

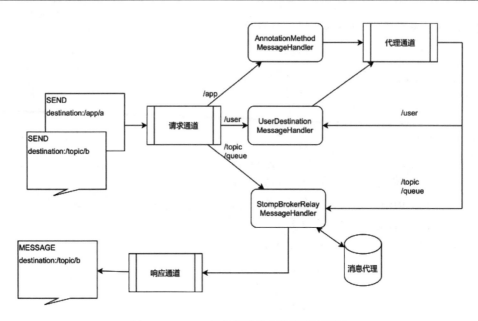

图 8.5 STOMP 服务端消息分发流程示意图

图中/app 等前缀均为默认目的地前缀，SEND 命令指向不同的前缀，响应的处理流程会不同。例如，目的地为/app 前缀的消息将会被路由至@MessageMapping 或@Subscri-beMapping 注解的方法中；以/topic 或/queue 为前缀的消息将会发送至 STOMP 代理中；以/user 为前缀的消息将会重定向至用户独有的目的地。

完成端点与前缀的注册后，可以着手编写消息以及消息控制器了。消息类实例/model/ChatMessage.java 如下：

```java
@Accessors(chain = true)
@Setter
@Getter
public class ChatMessage {
    private MessageType type;
    private String content;
    private String sender;

    public enum MessageType {
        CHAT,
        JOIN,
        LEAVE
    }
}
```

MessageType 用于表示消息类型，分别为：聊天信息、上线信息以及下线信息。

控制器示例/controller/ChatController.java 如下所示：

```java
@Controller
public class ChatController {

    @MessageMapping("/chat.sendMessage")
    @SendTo("/topic/public")
```

```
    public ChatMessage sendMessage(@Payload ChatMessage chatMessage) {
        return chatMessage;
    }

    @MessageMapping("/chat.addUser")
    @SendTo("/topic/public")
    public ChatMessage addUser(@Payload ChatMessage chatMessage,
                               SimpMessageHeaderAccessor headerAccessor) {
        headerAccessor.getSessionAttributes().put("username", chatMessage.getSender());
        return chatMessage;
    }

}
```

sendMessage 将会把消息转发到另外的目的地"/topic/public"。同样，addUser 将会把用户上线的信息转发到"/topic/public"。之后编写客户端的时候，客户端将会订阅"/topic/public"这一端点，最终实现消息的群发。

在实现用户上线通知之后，还可以实现下线通知。上线通知是通过服务端接收来自客户端的上线消息，之后再重定向至群发端点处。下线通知并不能如法炮制，因为 Web 端的下线往往异常断线居多，下线方面的处理需要在服务端完成。

下线通知可以基于 EventListener 实现，示例代码/controller/WebSocketEventListener.java 如下：

```
@Slf4j
@Component
public class WebSocketEventListener {

    @Autowired
    private SimpMessageSendingOperations messagingTemplate;

    @EventListener
    public void handleWebSocketConnectListener(SessionConnectedEvent event) {
        log.info("Received a new web socket connection");
    }

    @EventListener
    public void handleWebSocketDisconnectListener(SessionDisconnectEvent event) {
        StompHeaderAccessor headerAccessor = StompHeaderAccessor.wrap(event.getMessage());
        String username = (String) headerAccessor.getSessionAttributes().get("username");
        if (username != null) {
            log.info("User Disconnected : " + username);
            ChatMessage chatMessage = new ChatMessage();
            chatMessage.setType(ChatMessage.MessageType.LEAVE);
            chatMessage.setSender(username);
            messagingTemplate.convertAndSend("/topic/public", chatMessage);
        }
    }
}
```

这样的处理方式有些类似于使用 javax 提供的 WebSocket API。通过监听会话断开事件

SessionDisconnectEvent 获取用户信息，之后构造系统消息并通过 messagingTemplate 将消息转发至"/topic/public"端点。

2. 客户端部分

在该阶段，客户端需要实现基础的连接与消息发送。客户端主页文件/resource/static/index.html 如下：

```
01  <!DOCTYPE html>
02  <html>
03  <head>
04    <meta name="viewport" content="width=device-width, initial-scale=1.0, minimum-scale=1.0">
05    <title>Spring Boot WebSocket Chat Application</title>
06  </head>
07  <style>
08    .hidden {
09      display: none;
10    }
11
12  </style>
13  <body>
14  <noscript>
15    <h2>Sorry! Your browser doesn't support Javascript</h2>
16  </noscript>
17
18  <div id="username-page">
19    <div class="username-page-container">
20      <h1 class="title">Type your username</h1>
21      <form id="usernameForm" name="usernameForm">
22        <div class="form-group">
23          <input type="text" id="name" placeholder="Username" autocomplete="off" class="form-control"/>
24        </div>
25        <div class="form-group">
26          <button type="submit" class="accent username-submit">Start Chatting</button>
27        </div>
28      </form>
29    </div>
30  </div>
31
32  <div id="chat-page" class="hidden">
33    <div class="chat-container">
34      <div class="chat-header">
35        <h2>Spring WebSocket Chat Demo</h2>
36      </div>
37      <div class="connecting">
38        Connecting...
39      </div>
40      <ul id="messageArea">
41
42      </ul>
43      <form id="messageForm" name="messageForm" nameForm="messageForm">
```

```
44      <div class="form-group">
45        <div class="input-group clearfix">
46          <input type="text" id="message" placeholder="Type a message..." autocomplete="off"
47              class="form-control"/>
48          <button type="submit" class="primary">Send</button>
49        </div>
50      </div>
51    </form>
52  </div>
53 </div>
54
55 <script src="https://cdnjs.cloudflare.com/ajax/libs/sockjs-client/1.1.4/sockjs.min.js"></script>
56 <script src="https://cdnjs.cloudflare.com/ajax/libs/stomp.js/2.3.3/stomp.min.js"></script>
57 <script src="/js/main.js"></script>
58 </body>
59 </html>
```

代码说明：

- 18~30 行声明用于用户登录的输入框以及提交按钮。
- 32~53 行声明用于聊天信息展示以及输入。

对应的脚本文件/js/main.js 如下：

```
01 'use strict';
02
03 var usernamePage = document.querySelector('#username-page');
04 var chatPage = document.querySelector('#chat-page');
05 var usernameForm = document.querySelector('#usernameForm');
06 var messageForm = document.querySelector('#messageForm');
07 var messageInput = document.querySelector('#message');
08 var messageArea = document.querySelector('#messageArea');
09 var connectingElement = document.querySelector('.connecting');
10
11 var stompClient = null;
12 var username = null;
13 //连接函数
14 function connect(event) {
15   username = document.querySelector('#name').value.trim();
16   //输入用户名后，隐藏用户名输入框并建立 STOMP 连接
17   if(username) {
18     usernamePage.classList.add('hidden');
19     chatPage.classList.remove('hidden');
20     var socket = new SockJS('/ws');
21     stompClient = Stomp.over(socket);
22     stompClient.connect({}, onConnected, onError);
23   }
24   event.preventDefault();
25 }
26
```

```
27  //用于处理 STOMP 成功连接
28  function onConnected() {
……
45  }
46
47  //用于处理发送消息时间
48  function sendMessage(event) {
……
61  }
62
63  //收到消息后的处理
64  function onMessageReceived(payload) {
……
87  }
88
89  usernameForm.addEventListener('submit', connect, true)
90  messageForm.addEventListener('submit', sendMessage, true)
```

其中 onConnected 函数的实现如下：

```
28  function onConnected() {
29    // 订阅'/topic/public'
30    stompClient.subscribe('/topic/public', onMessageReceived);
31
32    // 发送用户名至'/app/chat.addUser'
33    stompClient.send("/app/chat.addUser",
34      {},
35      JSON.stringify({sender: username, type: 'JOIN'})
36    )
37
38    connectingElement.classList.add('hidden');
39  }
```

在连接成功之后，stompClient 订阅"/topic/public"端点，调用 onMessageReceived 函数对结果进行处理。同时将上线信息发送至"/app/chat.addUser'"端点。最后隐藏输入用户名的 connectingElement。

onMessageReceived 的实现如下：

```
64  function onMessageReceived(payload) {
65    var message = JSON.parse(payload.body);
66    var messageElement = document.createElement('li');
67    if(message.type === 'JOIN') {
68      messageElement.classList.add('event-message');
69      message.content = message.sender + ' joined!';
70    } else if (message.type === 'LEAVE') {
71      messageElement.classList.add('event-message');
72      message.content = message.sender + ' left!';
73    } else {
74      messageElement.classList.add('chat-message');
75      var usernameElement = document.createElement('span');
76      var usernameText = document.createTextNode((message.sender + ' :'));
77      usernameElement.appendChild(usernameText);
78      messageElement.appendChild(usernameElement);
79    }
80
```

```
81    var textElement = document.createElement('p');
82    var messageText = document.createTextNode(message.content);
83    textElement.appendChild(messageText);
84    messageElement.appendChild(textElement);
85    messageArea.appendChild(messageElement);
86    messageArea.scrollTop = messageArea.scrollHeight;
87  }
```

消息将会通过 JSON.parse 进行反序列化，之后根据消息的 type 属性判断消息类型，以不同的显示方式来处理不同类型的消息。

sendMessage 的实现如下：

```
48  function sendMessage(event) {
49    var messageContent = messageInput.value.trim();
50
51    if(messageContent && stompClient) {
52      var chatMessage = {
53        sender: username,
54        content: messageInput.value,
55        type: 'CHAT'
56      };
57      stompClient.send("/app/chat.sendMessage", {},
        JSON.stringify(chatMessage));
58      messageInput.value = '';
59    }
60    event.preventDefault();
61  }
```

需要发送的消息将会被构造成服务端中所定义的格式，通过 stompClient 发送至"/app/chat.sendMessage"端点。

以上是消息模块的内容，聊天服务程序的雏形在本小节便完成了。接下来将通过集成 Spring Security 与 MongoDB，为该程序加上存储模块与安全模块，从而提供除聊天功能之外的更多功能。

8.3.6 模块配置

模块配置方面主要针对安全模块进行，Spring Security 不仅需要过滤 HTTP 协议的请求，同样还需要保护 WebSocket 连接。

WebSocket 安全配置示例代码/config/SocketSecurityConfig.java：

```
@Configuration
public class SocketSecurityConfig extends
AbstractSecurityWebSocketMessageBrokerConfigurer {

    @Override
    protected void configureInbound(MessageSecurityMetadataSourceRegistry messages) {
        messages
            .simpDestMatchers("/**").authenticated()
            .anyMessage().authenticated();
    }
```

```
    @Override
    protected boolean sameOriginDisabled() {
        //关闭同源策略
        return true;
    }
}
```

该配置将会打开所有 WebSocket 端点的连接验证。

HTTP 安全配置示例代码/config/SecurityConfig.java：

```
@Configuration
@RequiredArgsConstructor
@EnableGlobalMethodSecurity(prePostEnabled = true, securedEnabled = true,
jsr250Enabled = true)
public class SecurityConfig extends WebSecurityConfigurerAdapter {

    private final ObjectMapper mapper;
    private final UserDetailsService userDetailsService;

    @Bean
    public PasswordEncoder bCryptPasswordEncoder() {
        return new BCryptPasswordEncoder();
    }

    @Override
    public void configure(WebSecurity web) {
        //将以下路径移出过滤链
        web.ignoring().antMatchers("/session", "/index.html", "/js/**");
    }

    @Override
    protected void configure(HttpSecurity http) throws Exception {
        ......
    }
}
```

其中 configure 方法实现如下：

```
@Override
protected void configure(HttpSecurity http) throws Exception {
    http
            //开启登录配置
            .authorizeRequests()
            //表示所有接口，登录之后就能访问
            .anyRequest().authenticated().and()
            .formLogin()
            .loginPage("/index.html")
            //登录处理接口
            .loginProcessingUrl("/login")
            //定义登录时，用户名的 key，默认为 username
            .usernameParameter("username")
            //定义登录时，用户密码的 key，默认为 password
            .passwordParameter("password")
            //登录成功的处理器
            .successHandler((req, resp, authentication) -> {
```

```java
            resp.setContentType("application/json;charset=utf-8");
            PrintWriter out = resp.getWriter();
            out.write(mapper.writeValueAsString(R.ok()));
            out.flush();
        })
        //登录失败的处理器
        .failureHandler((req, resp, exception) -> {
            resp.setContentType("application/json;charset=utf-8");
            PrintWriter out = resp.getWriter();
            out.write(mapper.writeValueAsString(R.failed("登录失败")));
            out.flush();
        })
        //和表单登录相关的接口可以不受限制直接通过
        .permitAll().and()
        .logout()
        .logoutUrl("/logout")
        //登出成功的处理程序
        .logoutSuccessHandler((req, resp, authentication) -> {
            resp.setContentType("application/json;charset=utf-8");
            PrintWriter out = resp.getWriter();
            out.write(mapper.writeValueAsString(R.ok()));
            out.flush();
        })
        .permitAll().and()
        .httpBasic().disable()
        .csrf().disable()
        .userDetailsService(userDetailsService);
}
```

8.3.7 注册登录

注册登录功能依赖于用户信息的读写，需要优先实现读写相关的功能。得益于 Spring Data 对 MongoDB 的支持，这部分代码非常简单。

1. 服务端部分

实体类/entity/TUser.java：

```java
@Document
@Accessors(chain = true)
@Setter
@Getter
public class TUser {

    @Id
    private String id;
    private String username;
    private String password;

}
```

数据访问类/repository/UserRepository.java：

```java
public interface UserRepository extends MongoRepository<TUser, String> {
```

}

用户注册控制器/controller/UserController.java：

```
@RestController
@RequiredArgsConstructor
public class UserController {

    private final UserRepository userRepository;
    private final PasswordEncoder encoder;

    @PostMapping("/session")
    public R<?> loginOrRegister(HttpServletRequest request, @RequestParam String username, @RequestParam String password) throws ServletException {
        Optional<TUser> userOptional = userRepository.findOne(Example.of(new TUser()
   .setUsername(username)));
        if (!userOptional.isPresent()) {
            TUser user = new TUser()
                    .setUsername(username)
                    .setPassword(encoder.encode(password));
            userRepository.insert(user);
        } else {
            return R.failed("账号已存在");
        }
        return R.ok();
    }

}
```

在用户注册的过程中，使用 PasswordEncoder 为用户的明文密码加密后再保存至数据库。

登录功能相关示例代码/security/AuthorityUserDetailServiceImpl.java：

```
@Transactional
@Service
@RequiredArgsConstructor
public class AuthorityUserDetailServiceImpl implements UserDetailsService {

    private final UserRepository userRepository;

    @Override
    public UserDetails loadUserByUsername(String username) throws UsernameNotFoundException {
        Optional<TUser> userOptional = userRepository.findOne(Example.of(new TUser().setUsername(username)));
        if (!userOptional.isPresent()) {
            throw new UsernameNotFoundException("username not found");
        }
        TUser user = userOptional.get();
        return new User(user.getUsername(), user.getPassword(),
Collections.singleton(new SimpleGrantedAuthority("USER")));
    }
}
```

登录过程中，Spring Security 将委托该类完成登录的业务逻辑。

2. 客户端部分

因为加入账号和密码登录的缘故，客户端的登录页面需要做相应调整。index.html 需要加入密码的输入框：

```
19  <div id="username-page">
20    <div class="username-page-container">
21      <h1 class="title">Type your username</h1>
22      <form id="usernameForm" name="usernameForm">
23        <div class="form-group">
24          <input type="text" id="name" placeholder="Username" autocomplete="off" class="form-control"/>
25        </div>
26        <div class="form-group">
27          <input type="password" id="password" placeholder="Password" autocomplete="off" class="form-control"/>
28        </div>
29        <div class="form-group">
30          <button type="submit" class="accent username-submit">Start Chatting</button>
31        </div>
32      </form>
33    </div>
34  </div>
```

main.js 部分将加入登录前自动注册的功能：

```
11  var stompClient = null;
12  var username = null;
13  var password = null;
14
15  function connect(event) {
16    username = document.querySelector('#name').value.trim();
17    password = document.querySelector('#password').value.trim();
18
19    if(username&&password) {
20      //登录前完成自动注册
21      loginWithRetry(username,password);
22    }
23    event.preventDefault();
24  }
```

loginWithRetry 函数以及相关函数实现如下：

```
25  function loginWithRetry(username,password){
26    //请求登录接口
27    var xhr = new XMLHttpRequest();
28    xhr.open('POST', '/login');
29    xhr.setRequestHeader('Content-Type','application/x-www-form-urlencoded');
30    xhr.send(`username=${username}&password=${password}`);
31    xhr.onload = () => {
32      var sessionResponse = JSON.parse(xhr.responseText);
33      if(sessionResponse&&sessionResponse.code==0){
34        //登录成功建立 STOMP 连接
35        stompConnect();
36      }else{
```

```
37              //登录失败则尝试注册
38              registerThenLogin(username,password);
39           }
40        }
41    }
42
43    function registerThenLogin(username,password){
44        //请求注册接口
45        var registerXhr = new XMLHttpRequest();
46        registerXhr.open('POST', '/session');
47        registerXhr.setRequestHeader('Content-Type',
          'application/x-www-form-urlencoded');
48        registerXhr.send(`username=${username}&password=${password}`);
49        registerXhr.onload = () => {
50           var sessionResponse = JSON.parse(registerXhr.responseText);
51           if(sessionResponse&&sessionResponse.code==0){
52              //注册成功后自动登录
53              var xhr = new XMLHttpRequest();
54              xhr.open('POST', '/login');
55              xhr.setRequestHeader('Content-Type',
                'application/x-www-form-urlencoded');
56              xhr.send(`username=${username}&password=${password}`);
57              xhr.onload = () => {
58                 var sessionResponse = JSON.parse(xhr.responseText);
59                 if(sessionResponse&&sessionResponse.code==0){
60                    stompConnect();
61                 }else{
62                    alert(sessionResponse.message);
63                 }
64              }
65           }else{
66              alert(sessionResponse.message);
67           }
68        }
69    }
```

8.3.8 聊天记录

聊天记录将会提供对聊天信息的记录功能，当用户登录进入聊天室后，可以收到登录前的聊天室内的记录。

1. 服务端部分

该功能需要修改/model/ChatMessage.java 为/entity/TChatMessage.java，以便聊天信息被记录到数据库。示例代码如下：

```
@Document
@Accessors(chain = true)
@Setter
@Getter
public class TChatMessage {
    @Id
    private String id;
```

```
    private MessageType type;
    private String content;
    private String sender;
    private LocalDateTime createTime;

    public enum MessageType {
        CHAT,
        JOIN,
        LEAVE
    }
}
```

数据访问类/repository/ChatMessageRepository.java：

```
    public interface ChatMessageRepository extends MongoRepository<TChatMessage,
String> {

        List<TChatMessage> findTop10ByOrderByCreateTimeDesc();

    }
```

ChatMessageRepository 根据 Spring Data 语法，提供一个根据消息创建时间查询数据库内最后 10 条历史记录的抽象方法。Spring Data MongoDB 将会在程序启动后为该抽象方法生成具体实现。

调整后的 ChatController.java：

```
    @Slf4j
    @Controller
    @RequiredArgsConstructor
    public class ChatController {

        private final ChatMessageRepository chatMessageRepository;
        private final UserRepository userRepository;
        private final SimpMessagingTemplate messagingTemplate;

        @MessageMapping("/chat.sendMessage")
        @SendTo("/topic/public")
        public TChatMessage sendMessage(@Payload TChatMessage chatMessage) {
            //在记录中留下类型为CHAT的信息
            if (chatMessage.getType() == TChatMessage.MessageType.CHAT) {
                chatMessageRepository.save(chatMessage.setCreateTime
(LocalDateTime.now()));
            }
            return chatMessage;
        }

        @MessageMapping("/chat.addUser")
        @SendTo("/topic/public")
        public TChatMessage addUser(@Payload TChatMessage chatMessage,
SimpMessageHeaderAccessor headerAccessor) {
            headerAccessor.getSessionAttributes().put("username",
chatMessage.getSender());
            return chatMessage;
        }

        @SubscribeMapping("/chat.lastTenMessage")
```

```java
    public List<TChatMessage> addUser() {
        //读取10条历史记录
        List<TChatMessage> ret = chatMessageRepository.findTop10ByOrderByCreateTimeDesc();
        ret.sort(Comparator.comparing(TChatMessage::getCreateTime));
        return ret;
    }
}
```

2. 客户端部分

客户端部分需要在建立 STOMP 连接时订阅历史记录的主题，以便接收对应的消息。main.js 的改动部分如下：

```javascript
84  /*STOMP 连接成功*/
85  function onConnected() {
86    //订阅群聊主题
87    stompClient.subscribe('/topic/public', onMessageReceived);
88    //订阅获取历史记录的主题
89    stompClient.subscribe('/app/chat.lastTenMessage', onMessageReceived);
90    //发送加入群聊的消息
91    stompClient.send("/app/chat.addUser",
92      {},
93      JSON.stringify({sender: username, type: 'JOIN'})
94    )
95    connectingElement.classList.add('hidden');
96  }
97
......
118
119  /*处理来自服务端的消息*/
120  function onMessageReceived(payload) {
121    var body = JSON.parse(payload.body);
122    //适配不同类型的返回结果。如果非数组则转为数组
123    var message = body instanceof Array? body:[body];
124    for(var i in message){
125      var messageElement = document.createElement('li');
126      if(message[i].type === 'JOIN') {
127        //处理加入群聊信息
128        messageElement.classList.add('event-message');
129        message[i].content = message[i].sender + ' joined!';
130        //处理离开群聊信息
131      } else if (message[i].type === 'LEAVE') {
132        messageElement.classList.add('event-message');
133        message[i].content = message[i].sender + ' left!';
134      } else {
135        //处理聊天信息
136        messageElement.classList.add('chat-message');
137        var usernameElement = document.createElement('span');
138        var usernameText = document.createTextNode((message[i].sender + ' :'));
139        usernameElement.appendChild(usernameText);
140        messageElement.appendChild(usernameElement);
141      }
142      var textElement = document.createElement('p');
143      var messageText = document.createTextNode(message[i].content);
```

```
144        textElement.appendChild(messageText);
145        messageElement.appendChild(textElement);
146        messageArea.appendChild(messageElement);
147        messageArea.scrollTop = messageArea.scrollHeight;
148    }
149 }
```

8.3.9 私聊功能

实现了群聊功能之后，还可以通过 SimpMessagingTemplate 实现私聊功能。私聊功能需要客户端在订阅公共话题之外，再另外订阅各自的私聊话题。服务端根据消息内容对消息做路由转发。

1. 服务端部分

私聊功能要在/topic 之外另外启用私聊的路径/user。修改后的 WebSocketConfig.java：

```
@Configuration
@EnableWebSocketMessageBroker
public class WebSocketConfig implements WebSocketMessageBrokerConfigurer {

    @Override
    public void registerStompEndpoints(StompEndpointRegistry registry) {
        //注册 STOMP 端点
        registry.addEndpoint("/ws").withSockJS();
    }

    @Override
    public void configureMessageBroker(MessageBrokerRegistry registry) {
        //配置 app 与 topic 端点
        registry.setApplicationDestinationPrefixes("/app");
        registry.enableSimpleBroker("/topic", "/user");
    }
}
```

为了路由转发能在服务端成功实现，需要将消息的结构进行扩展，在发信人之外新增收信人属性。修改后的 TChatMessage.java：

```
@Document
@Accessors(chain = true)
@Setter
@Getter
public class TChatMessage {
    @Id
    private String id;
    /**
     * 消息类型
     */
    private MessageType type;
    /**
     * 消息内容
     */
    private String content;
    /**
     * 发信人
```

```java
     */
    private String sender;
    /**
     * 收信人
     */
    private String receiver;
    private LocalDateTime createTime;

    public enum MessageType {
        CHAT,
        JOIN,
        LEAVE
    }
}
```

对消息的路由需要在控制器中实现，并且历史记录相关的部分也需要做一定调整。修改后的 ChatController.java：

```java
    public class ChatController {

    ……

        @MessageMapping("/chat.sendMessage")
        public void sendMessage(Principal principal, @Payload TChatMessage chatMessage) {
            //在记录中留下类型为 CHAT 的信息
            if (chatMessage.getType() == TChatMessage.MessageType.CHAT) {
                chatMessageRepository.save(chatMessage.setCreateTime(LocalDateTime.now()));
            }
            if (StringUtils.isEmpty(chatMessage.getReceiver())) {
                //根据收信人进行路由，没有该属性则视作群发消息
                messagingTemplate.convertAndSend("/topic/public", chatMessage);
            } else {
                //回传给发信人
                messagingTemplate.convertAndSendToUser(chatMessage.getSender(), "/notification", chatMessage);
                //发送给收信人
                messagingTemplate.convertAndSendToUser(chatMessage.getReceiver(), "/notification", chatMessage);
            }

        }

    ……

        @SubscribeMapping("/chat.lastTenMessage")
        public List<TChatMessage> addUser(Principal principal) {
            //读取 10 条历史记录
            Query query = new Query();
            //查询群聊、自己发送或者发送给自己的历史记录
            Criteria criteria = new Criteria().orOperator(
                    Criteria.where("receiver").is(null),
                    Criteria.where("sender").is(principal.getName()),
```

```
                Criteria.where("receiver").is(principal.getName()));
        query.addCriteria(criteria)
                //按时间倒序
                .with(Sort.by(Sort.Direction.DESC, "createTime"))
                .limit(10);
        List<TChatMessage> ret = mongoTemplate.find(query,
TChatMessage.class);
        ret.sort(Comparator.comparing(TChatMessage::getCreateTime));
        return ret;
    }
}
```

2. 客户端部分

因为私聊功能加入了收信人属性,客户端需要加入指定收信人的部分。修改后的 index.html:

```
......
07  <style>
08  .hidden {
09    display: none !important;
10  }
11
12  .send-to {
13    display:inline-block;
14    color:gray;
15    margin:0;
16  }
17  </style>
......
51      <form id="messageForm" name="messageForm" nameForm="messageForm">
52        <div class="form-group">
53          <div class="input-group clearfix">
54            <div class="send-to hidden">
55              <p class="send-to">To</p>
56              <p class="send-to receiver"></p>
57              <p class="send-to">:</p>
58            </div>
59            <input type="text" id="message" placeholder="Type a message..." autocomplete="off"
60                class="form-control"/>
61            <button type="submit" class="primary">Send</button>
62          </div>
63        </div>
64      </form>
65    </div>
66  </div>
......
```

index.html 中加入 sent-to 类以及显示收信人信息的标签（代码 54~58 行）。main.js 中 STOMP 连接、发送服务端以及收信等部分都需要进行一些修改。main.js 中 STOMP 连接的相关改动:

```
01  'use strict';
02
03  var usernamePage = document.querySelector('#username-page');
04  var chatPage = document.querySelector('#chat-page');
05  var usernameForm = document.querySelector('#usernameForm');
```

```
06    var messageForm = document.querySelector('#messageForm');
07    var messageInput = document.querySelector('#message');
08    var messageArea = document.querySelector('#messageArea');
09    var connectingElement = document.querySelector('.connecting');
10    var sendToElement = document.querySelector('.send-to');
11    var receiverElement = document.querySelector('.receiver');
12
13    var stompClient = null;
14    var username = null;
15    var password = null;
16    var receiver = null;
……
100   /*STOMP 连接成功*/
101   function onConnected() {
102     //订阅群聊主题
103     stompClient.subscribe('/topic/public', onMessageReceived);
104     //订阅私聊主题
105     stompClient.subscribe(` / user / $ {
106       username
107     }
108   /notification`, onMessageReceived);
109     //订阅获取历史记录的主题 stompClient.subscribe('/app/chat.lastTenMessage',
 onMessageReceived);
110     //发送加入群聊的消息
111     stompClient.send("/app/chat.addUser", {},
112   JSON.stringify({
113     sender: username,
114     type: 'JOIN'
115   })) connectingElement.classList.add('hidden');
116   }
```

建立 STOMP 连接过程中，需要另外订阅私聊主题。向服务端发送聊天消息的相关改动：

```
125   function sendMessage(event) {
126     var messageContent = messageInput.value.trim();
127     if (messageContent && stompClient) {
128       var chatMessage = {
129         //新增收信人属性
130         receiver: receiver,
131         sender: username,
132         content: messageInput.value,
133         type: 'CHAT'
134       };
135       stompClient.send("/app/chat.sendMessage", {},
136   JSON.stringify(chatMessage));
137       messageInput.value = '';
138     }
139     event.preventDefault();
140   }
```

main.js 中处理来自服务端的消息：

```
142   function onMessageReceived(payload) {
143     var body = JSON.parse(payload.body);
144     //适配不同类型的返回结果。如果非数组则转为数组
145     var message = body instanceof Array ? body: [body];
```

```
146   for (var i in message) {
147     var messageElement = document.createElement('li');
148     if (message[i].type === 'JOIN') {
149       //处理加入群聊信息
150       messageElement.classList.add('event-message');
151       message[i].content = message[i].sender + ' joined!';
152       //处理离开群聊信息
153     } else if (message[i].type === 'LEAVE') {
154       messageElement.classList.add('event-message');
155       message[i].content = message[i].sender + ' left!';
156     } else {
157       //处理聊天信息
158       messageElement.classList.add('chat-message');
159       var usernameElement = document.createElement('span');
160       var usernameText = document.createTextNode((message[i].sender + ' :'));
161       usernameElement.appendChild(usernameText);
162       messageElement.appendChild(usernameElement);
163     }
164     var textElement = document.createElement('p');
165     var messageText = document.createTextNode(message[i].content);
166     textElement.appendChild(messageText);
167     messageElement.appendChild(textElement);
168     messageArea.appendChild(messageElement);
169     messageArea.scrollTop = messageArea.scrollHeight;
170     //监听消息展示区内的用户名
171     if (message[i].sender != username) {
172       messageElement.addEventListener('click',
173         function(e) {
174 selectOrCancelReceiver(e.target.parentElement.getAttribute("sender"));
175         },
176         true);
177     }
178   }
179 }
```

为当前在聊天室内活跃用户的用户名添加单击事件监听，单击过后调用选择或取消收信人的函数。该函数的具体实现如下：

```
180 /*选择或者取消收信人*/
181 function selectOrCancelReceiver(o) {
182   if (receiver == o) {
183     cancelReceiver();
184   } else {
185     receiver = o;
186     sendToElement.classList.remove('hidden');
187   }
188   receiverElement.innerHTML = receiver;
189
190 }
191 /*取消收信人*/
192 function cancelReceiver() {
193   receiver = null;
194   sendToElement.classList.add('hidden');
195 }
```

8.4 测试与验证

本节将通过集成测试与手工测试的方式来对程序功能进行验证。集成测试用于检验控制器的行为是否符合预期，能否返回期望的结果。手工测试用于验证客户端与服务端的集成结果是否满足需求设计中的要求。

8.4.1 集成测试

集成测试的重点在于测试消息模块。对消息模块的测试也基于 STOMP 协议，因此需要引入 WebSocketStompClient。示例代码/test/java/com/example.chat/ChatApplicationTests.java 如下：

```java
@SpringBootTest(classes = {ChatTestApplication.class}, webEnvironment = SpringBootTest.WebEnvironment.RANDOM_PORT)
@Import(SocketSecurityTestConfig.class)
class ChatApplicationTests {

    @LocalServerPort
    private Integer port;

    @Autowired
    private ObjectMapper objectMapper;

    private WebSocketStompClient webSocketStompClient;

    @BeforeEach
    public void setup() {
        this.webSocketStompClient = new WebSocketStompClient(new SockJsClient(
                Collections.singletonList(new WebSocketTransport(new StandardWebSocketClient()))));
        this.webSocketStompClient.setMessageConverter(new MappingJackson2MessageConverter());
    }

    @Test
    public void subscribePublic_thenSend() throws InterruptedException, ExecutionException, TimeoutException {
        //todo
    }

    @Test
    public void subscribePrivate_thenSend() throws InterruptedException, ExecutionException, TimeoutException {
        //todo
    }

    private String getWsPath() {
        return String.format("ws://localhost:%d/ws", port);
```

subscribePublic_thenSend 方法用于检验群聊接口，具体实现如下：

```java
@Test
public void subscribePublic_thenSend() throws InterruptedException,
ExecutionException, TimeoutException {
    BlockingQueue<TChatMessage> blockingQueue = new ArrayBlockingQueue(1);

    StompSession session = webSocketStompClient
            .connect(getWsPath(), new StompSessionHandlerAdapter() {
            })
            .get(1, SECONDS);
    session.subscribe("/topic/public", new StompFrameHandler() {
      @Override
      public Type getPayloadType(StompHeaders headers) {
        return TChatMessage.class;
      }

      @Override
      public void handleFrame(StompHeaders headers, Object payload) {
        blockingQueue.add((TChatMessage) payload);
      }
    });
    session.send("/app/chat.sendMessage", new TChatMessage()
            .setContent("hello")
            .setSender("someone")
            .setType(TChatMessage.MessageType.CHAT));

    TChatMessage ret = blockingQueue.poll(1, SECONDS);
    assertEquals("hello", ret.getContent());
}
```

因为该测试是一个异步的过程，所以需要一个支持等待元素的数据类型 BlockingQueue。在创建 StompSession 后对 "/topic/public" 进行订阅。订阅到对应端点后，一旦收到 STOMP 帧，handleFrame 将会被调用，并将收到的消息写入 blockingQueue 中。订阅完成后，向 "/app/chat.sendMessage" 发送消息。预测的结果是在 blockingQueue 中也会收到相同内容的消息。

对私聊的集成测试逻辑与群聊类似，具体实现如下：

```java
@Test
public void subscribePrivate_thenSend() throws InterruptedException,
ExecutionException, TimeoutException {
    BlockingQueue<TChatMessage> blockingQueue = new ArrayBlockingQueue(1);
    StompSession session = webSocketStompClient
            .connect(getWsPath(), new StompSessionHandlerAdapter() {
            })
            .get(1, SECONDS);
    session.subscribe(String.format("/user/%s/notification", "someone"),
  new StompFrameHandler() {
      @Override
        public Type getPayloadType(StompHeaders headers) {
          return TChatMessage.class;
        }
```

```
    @Override
    public void handleFrame(StompHeaders headers, Object payload) {
        blockingQueue.add((TChatMessage) payload);
    }
});
session.send("/app/chat.sendMessage", new TChatMessage()
        .setContent("hello")
        .setSender("someone")
        .setReceiver("anyone")
        .setType(TChatMessage.MessageType.CHAT));

TChatMessage ret = blockingQueue.poll(1, SECONDS);
assertEquals("hello", ret.getContent());
}
```

因为系统处于 Spring Security 的保护下，但是在测试过程中该框架将会带来不必要的麻烦。为了避免 Spring Security 在测试过程中带来的复杂度，本示例中需要将其屏蔽。具体方法是排除 Spring Security 的自动配置，并且另外声明一个测试配置覆盖掉 SocketSecurityConfig。

ChatApplicationTests 中声明的启动类 ChatTestApplication.java 如下：

```
@SpringBootApplication(exclude = {SecurityAutoConfiguration.class})
public class ChatTestApplication {

    public static void main(String[] args) {
        SpringApplication.run(ChatApplication.class, args);
    }

}
```

ChatApplicationTests 中另外引入的测试配置如下：

```
@TestConfiguration
public class SocketSecurityTestConfig extends
AbstractSecurityWebSocketMessageBrokerConfigurer {

    @Override
    protected void configureInbound(MessageSecurityMetadataSourceRegistry messages) {
        //do nothing
    }

    @Override
    protected boolean sameOriginDisabled() {
        return true;
    }
}
```

8.4.2 手工测试

手工测试将根据功能实现情况，在不同的阶段进行。

1. 消息模块

启动服务器并访问主页地址，默认地址 http://localhost:8080/，页面如图 8.6 所示。

> **注 意**
>
> 测试消息模块时，可以在依赖文件中暂时屏蔽 spring-boot-starter-security 模块。

图 8.6 初版登录页面

输入用户名之后，单击 Start Chatting 按钮即可进入聊天页面，如图 8.7 所示。

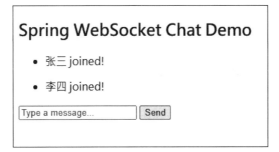

图 8.7 聊天页面

当任意一端在输入框内输入消息，并单击 Send 按钮发送，消息都将会展示在展示框中，如图 8.8 所示。

图 8.8 聊天示意图

关闭其中一方的页面，下线信息将群发至公屏，如图 8.9 所示。

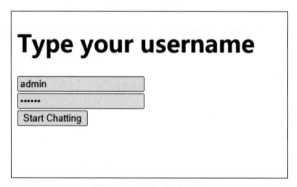

图 8.9　下线示意图

2. 注册登录

该阶段将验证用户是否可以输入账号和密码后自动完成注册并登录，用户登录页面如图 8.10 所示。

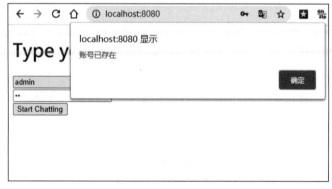

图 8.10　用户登录页面

登录成功后将跳转至聊天页面。如果输入了错误的密码，将弹框提示，如图 8.11 所示。

图 8.11　登录失败提示

3. 聊天记录

聊天记录功能将允许新加入的用户获取 10 条历史记录。发送超过 10 条聊天信息，发送端内容如图 8.12 所示。

接收端将按顺序收到 10 条历史记录，如图 8.13 所示。

图 8.12　发送端内容　　　　　图 8.13　接收端内容

4. 私聊功能

私聊功能将允许单击消息展示框内的用户名来指定私聊对象（见图 8.14），并且对私聊对象发送私信。

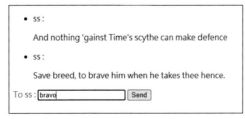

图 8.14　用户 fz 选择私聊对象

私信内容仅对发送方与接收方可见，对其余用户不可见，如图 8.15~图 8.17 所示。

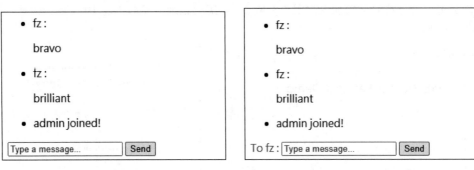

图 8.15　用户 fz 发送内容　　　　图 8.16　用户 ss 接收内容

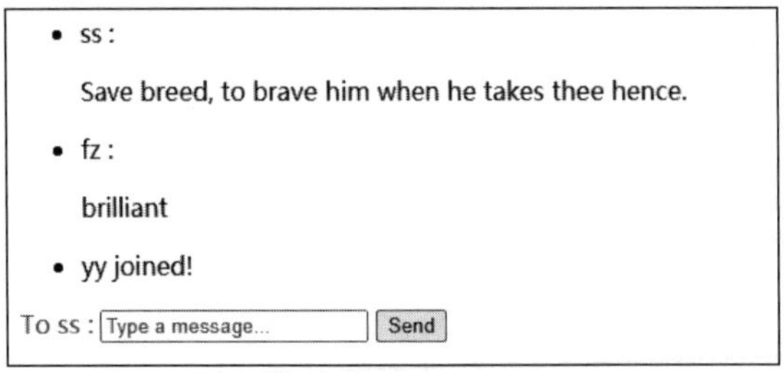

图 8.17　用户 yy 接收内容

第 9 章

实战 2：在线商城

本章将基于 Spring Boot 与模板引擎 Thymeleaf 构建一个在线商城程序，在实现过程中将会使用到之前章节中所介绍的大部分技术，巩固一下相应的知识点。

本章内容包含程序架构设计、数据库设计、各模块的实现以及程序测试，涉及的知识点如下：

- 基于 Spring Boot 的项目构建
- 基于 Thymeleaf 模板引擎的模板开发
- Spring Bean 作用域

9.1 架构设计

在线商城程序分为五个功能模块，分别对应五个页面或者页面片段。页面功能通过 Spring Boot 结合 Thymeleaf 实现；程序的持久层框架使用 JPA；安全框架依然使用 Spring Security。功能模块示意图如图 9.1 所示。

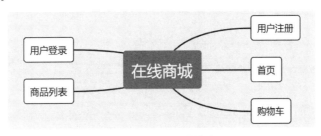

图 9.1　在线商城模块

9.2 框架搭建

框架搭建阶段，需要根据框架设计内容分析项目所需的依赖。根据依赖清单使用 Spring Initializr 初始化工程。在线商城为网站服务，需要引入 Web 依赖；数据存储依赖于关系型数据库，需要引入 JPA 以及对应数据库的驱动；注册与登录需要由安全框架实现，需要引入 Spring Security 相关依赖；页面使用 Thymeleaf 实现，需要引入 Thymeleaf 相关依赖。pom.xml 中主要依赖项如下：

```xml
<!--Web-->
<dependency>
    <groupId>org.springframework.boot</groupId>
    <artifactId>spring-boot-starter-web</artifactId>
</dependency>
<!--JPA-->
<dependency>
    <groupId>org.springframework.boot</groupId>
    <artifactId>spring-boot-starter-data-jpa</artifactId>
</dependency>
<!--验证相关依赖-->
<dependency>
    <groupId>org.springframework.boot</groupId>
    <artifactId>spring-boot-starter-validation</artifactId>
</dependency>
<!--安全框架-->
<dependency>
    <groupId>org.springframework.boot</groupId>
    <artifactId>spring-boot-starter-security</artifactId>
</dependency>
<!--H2 数据库-->
<dependency>
    <groupId>com.h2database</groupId>
    <artifactId>h2</artifactId>
    <scope>runtime</scope>
</dependency>
<!--Thymeleaf 相关依赖-->
<dependency>
    <groupId>org.springframework.boot</groupId>
    <artifactId>spring-boot-starter-thymeleaf</artifactId>
</dependency>
<dependency>
    <groupId>org.thymeleaf.extras</groupId>
    <artifactId>thymeleaf-extras-springsecurity5</artifactId>
</dependency>
```

在线商城的项目包的基本结构如图 9.2 所示。

```
src
├main
│ ├java
│ │ └com
│ │    └example
│ │       └cart
│ │          │ CartApplication.java
│ │          ├config
│ │          ├controller
│ │          ├exception
│ │          ├model
│ │          ├repository
│ │          ├service
│ │          └util
│ │
│ └resources
│    │ application.properties
│    ├sql
│    ├static
│    └templates
└test
```

图 9.2　项目包结构

结构分明的项目结构对程序的编写与维护是有益的，该程序项目的包结构说明如下：

- config 包含项目的配置文件
- controller 包含控制器文件
- exception 包含自定义异常类
- model 包含模型对象
- repository 包含数据访问层文件
- service 包含服务类
- util 包含工具类
- resources 包含静态文件、模板以及项目初始化所需的 SQL 语句

9.3　数据库设计

该程序所使用的数据库为关系型数据库，在功能实现之前应当根据需求对数据库结构进行设计。对需求进行简单的分析之后可知需要持久化的数据有：用户信息、权限信息、产品信息以及订单信息。数据库 E-R 图如图 9.3 所示。

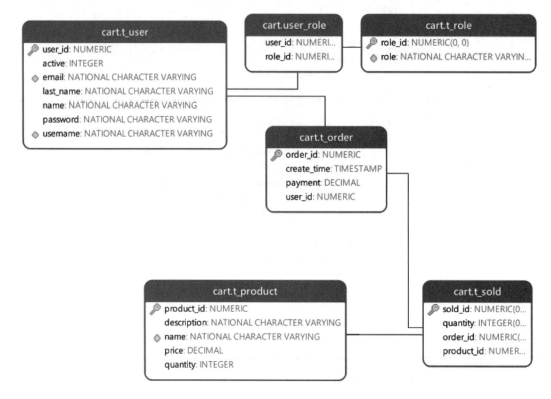

图 9.3　在线商城 E-R 图

与实体对应的表名以"t_"作前缀，以避免表名与数据库关键字重名（如订单表 order）。表说明如下：

- t_user：用户信息表，用于存储用户的登录信息以及基本信息。
- t_role：权限表，用于存储权限信息。
- user_role：中间表，用以支持 t_user 与 t_role 之间的多对多关系的映射。
- t_order：订单表，用以存储订单信息。订单表中包含售出时间、售出记录。用户表与订单表为一对多的关系（一位用户拥有多份订单）。
- t_sold：售出记录表，用以存储售出商品记录。订单表与售出记录表为一对多的关系（一份订单内包含多条售出记录）。
- t_product：产品表，用以存储产品信息。产品表与售出记录表为一对多的关系（一种产品对应多条售出记录）。

9.4　功能实现

为实现程序的功能，需要首先完成页面。根据页面展示以及功能需求编写每个功能对应的控制器。为了更好地完成页面编写的任务，在开始实现程序之前掌握一个页面模板引擎的使用会是一个好的选择。

9.4.1 模板引擎 Thymeleaf

该程序的实现涉及不少 UI 页面的编写，性能强大并且容易上手的技术可以为该环节省下不少精力。本示例中将使用 Thymeleaf 配合 Spring Boot 的 Web 模块实现展示层。

Thymeleaf 是一个现代化的服务端模板引擎，可以用于处理 HTML、XML、JavaScript、CSS 这些类型的文件。在本示例中，Thymeleaf 主要用于开发展示层所需的 HTML 模板。Thymeleaf 是 Spring 官方所推荐使用的模板引擎。Thymeleaf 具有很强的扩展性，它的模板功能可以确保没有后端服务的支持也能对模板进行原型制作，相对其他的模板引擎更灵活与高效。

Spring Boot 提供了针对 Thymeleaf 的自动配置，引入对应版本的 spring-boot-starter-thymeleaf 即可开始使用 Thymeleaf。

1. 简单示例

Thymeleaf 结合 Spring Boot 进行开发非常简便，以一个 HelloWorld 程序为例。程序向模板传递变量"text"，模板将变量的内容在页面上打印出来。Spring Boot 控制器示例 HelloWorldController.java 如下，该示例示范了使用 ModelAndView 以及 String 这两种返回类型指定视图文件的方式：

```java
@Controller
public class HelloWorldController {

    @GetMapping("/hello-world")
    public ModelAndView helloWorld() {
        ModelAndView modelAndView = new ModelAndView();
        modelAndView.addObject("text", "hello world! ");
        modelAndView.setViewName("/hello-world");
        return modelAndView;
    }

    @GetMapping("/greeting")
    public String greeting(Model model) {
        model.addAttribute("text", "你好");
        return "/greeting";
    }

}
```

为了完成对变量的展示，需要针对不同的方法编写对应的视图文件。在 resources/templates 路径下分别创建 hello-world.html 以及 greeting.html，hello-world.html 内容如下：

```html
<!DOCTYPE HTML>
<html xmlns:th="http://www.thymeleaf.org">
<head>
    <meta charset="UTF-8">
    <title th:text="${text}"></title>
</head>
<body>
```

```
<h1 th:text="${text}" ></h1>
</body>
</html>
```

greeting.html 内容与 hello-world.html 内容相同，故不再说明。访问对应路径后，页面将会把变量的内容写入到模板当中并返回给浏览器。访问结果如图 9.4 所示。

图 9.4　Thymeleaf Hello World 结果图

2. 标准表达语法

为了掌握 Thymeleaf，需要了解 Thymeleaf 的标准表达式的语法。Thymeleaf 的标准表达式分为简单表达式、文字、文字操作等。标准表达式的摘要如下：

（1）简单表达式
- 变量表达式：${...}
- 选择变量表达式：*{...}
- 消息表达：#{...}
- 链接 URL 表达式：@{...}

（2）文字
- 文本文字：'one text', 'Another one!', ...
- 号码文字：0, 34, 3.0, 12.3, ...
- 布尔文字：true, false
- 空文字：null
- 文字标记：one, sometext, main, ...

（3）文字操作
- 字符串串联：+

- 文字替换：|The name is ${name}|

（4）算术运算

- 二元运算符：+，-，*，/，%
- 减号（一元运算符）：-

（5）布尔运算

- 二元运算符：and，or
- 布尔否定（一元运算符）：!，not

（6）比较符运算

- 比较：>, <, >=, <= (gt, lt, ge, le)
- 等号运算符：==, != (eq, ne)

（7）条件运算

- 如果-则： (if) ? (then)
- 如果-则-否则： (if) ? (then) : (else)
- 默认： (value) ?: (defaultvalue)

通过以上语法构造的表达式将可以应用到 Thymeleaf 所提供的各种功能中。

3. 设置属性值

在第一个 Thymeleaf 的示例中可以观察到，th:text 属性提供了改变标签内容的功能。如果需要改变标签的属性值，可以通过 th:attr 实现。例如：

```
<textarea placeholder="提示内容……" th:attr="placeholder=${hin}" >
</textarea>
```

该语句的执行结果将会是：

```
<textarea placeholder="输入字符不要超过 20 个"></textarea>
```

这种设置属性值的方式适用性很广，包括标准属性和自定义属性。如果需要改变的属性为 HTML5 中的标准属性，可以使用更简洁的针对标准 HTML5 属性的 th 属性去实现。例如：

```
<textarea placeholder="提示内容……" th: placeholder "${hin}" ></textarea>
```

类似的 th 属性还包括"th:src""th:value""th:href"，等等。

4. 迭代

以上示例中的变量都是单个字符串。如果要处理一个集合类型的变量，则需要对变量的内容进行迭代。Thymeleaf 提供了 th:each 属性以帮助实现这类操作。th:each 的语法与 Java 中的 foreach 类似，例如：

```
    <table>
      <tr>
        <th>产品名</th>
```

```
        <th>价格</th>
        <th>数量</th>
    </tr>
    <tr th:each="p : ${products}">
        <td th:text="${p.name}"> Bighead carp</td>
        <td th:text="${p.price}">74.70</td>
        <td th:text="${p.quantity}">10</td>
    </tr>
</table>
```

5. 条件评估

Thymeleaf 提供了一种通过条件表达式控制模板行为的方式——th:if 与 th:unless。例如，目前有一批商品，当商品存在评论时才在页面中展示对应的链接标签。实现该功能的示例如下：

```
<table>
    <tr>
        <th>产品名</th>
        <th>价格</th>
        <th>数量</th>
    </tr>
    <tr th:each="p : ${products}">
        <td th:text="${p.name}"> Bighead carp</td>
        <td th:text="${p.price}">74.70</td>
        <td th:text="${p.quantity}">10</td>
        <td>
          <span th:text="${#lists.size(prod.comments)}">2</span> comment/s
          <a href="comments.html"
             th:href="@{/product/comments(prodId=${prod.id})}"
             th:if="${not #lists.isEmpty(prod.comments)}">view</a>
        </td>
    </tr>
</table>
```

该语句也可以用 th:unless 代替 th:if 和 not 的组合：

```
<a href="comments.html"
   th:href="@{/product/comments(prodId=${prod.id})}"
   th:unless ="${#lists.isEmpty(prod.comments)}">view</a>
```

6. 模板片段

在模板编写的过程中可能会有多个模板存在重复的部分。遇到这种情形，可以使用模板片段的功能，将重复的部分抽取出来，成为公用的模板片段。比如一组模板中共用一个页脚片段，可以使用 th:fragment 将对应的部分声明成模板片段，示例 footer.html 如下：

```
<html xmlns="http://www.w3.org/1999/xhtml"
xmlns:th="http://www.thymeleaf.org"
xmlns:sec="http://www.thymeleaf.org/thymeleaf-extras-springsecurity5">

    <head>
    </head>

    <body>
```

```html
<div th:fragment="footer">
    <script type="text/javascript"
src="webjars/bootstrap/3.3.7/js/bootstrap.min.js"></script>
</div>

</body>
</html>
```

在需要引用的模板中，使用 th:include 或者 th:replace 引入片段，例如：

```html
<div th:replace="/fragments/footer :: footer"/>
```

其中/fragments/footer 为模板路径，后面的 footer 为 th:fragment 中声明的片段名。

7. 局部变量

使用 th:with 可以使用局部变量功能。该功能与 Java 中的局部变量很相似，声明了局部变量之后，可以在声明变量的标签以内，通过变量名调用变量对应的值，例如：

```html
<div th:with="first=${guest[0]}">
  <p>
    第一位顾客的姓名是：<span th:text="${ first.name}">Xiao Jiang</span>.
  </p>
</div>
```

在对 Thymeleaf 有了大概的了解之后，可以开始着手编写功能页面以及后端服务了。

9.4.2 实体类

基于数据库设计，编写程序所需的实体类。用户类 User.java：

```java
@Accessors(chain = true)
@Setter
@Getter
@Entity
@Table(name = "t_user")
public class User implements Serializable {

    private static final long serialVersionUID = 57524743425783732l3L;

    @Id
    @GeneratedValue(strategy = GenerationType.IDENTITY)
    @Column(name = "user_id")
    private Long id;

    @Column(name = "email", unique = true, nullable = false)
    @Email(message = "*请提供有效的电子邮件")
    @NotEmpty(message = "*请提供电子邮件")
    private String email;

    @Column(name = "password", nullable = false)
    @Length(min = 5, message = "*您的密码必须至少包含 5 个字符")
    @NotEmpty(message = "*请提供您的密码")
    @JsonIgnore
    private String password;
```

```java
    @Column(name = "username", nullable = false, unique = true)
    @Length(min = 5, message = "*您的用户名必须至少包含 5 个字符")
    @NotEmpty(message = "*请提供您的名字")
    private String username;

    @Column(name = "name")
    @NotEmpty(message = "*请提供您的名字")
    private String name;

    @Column(name = "last_name")
    @NotEmpty(message = "*请提供您的姓氏")
    private String lastName;

    @Column(name = "active", nullable = false)
    private int active;

    @ManyToMany(cascade = CascadeType.ALL)
    @JoinTable(name = "user_role", joinColumns = @JoinColumn(name = "user_id"),
inverseJoinColumns = @JoinColumn(name = "role_id"))
    private Collection<Role> roles;

    @OneToMany(fetch = FetchType.LAZY, mappedBy = "user")
    private List<Order> orderList;

}
```

用户类与权限类为多对多的关系，因此需要使用到注释@ManyToMany。另外，用户类与订单类为一对多的关系，需要使用到注释@OneToMany。权限类 Role.java：

```java
@Accessors(chain = true)
@Setter
@Getter
@Entity
@Table(name = "t_role")
public class Role implements Serializable {

    private static final long serialVersionUID = -6969646367825535431L;

    @Id
    @GeneratedValue(strategy = GenerationType.IDENTITY)
    @Column(name = "role_id")
    private Long id;

    @Column(name = "role", unique = true)
    private String role;

    @ManyToMany(cascade = CascadeType.ALL, mappedBy = "roles")
    private Collection<User> users;

}
```

订单类 Order.java：

```java
@Accessors(chain = true)
```

```java
@Setter
@Getter
@Entity
@Table(name = "t_order")
public class Order implements Serializable {

    private static final long serialVersionUID = 7798851301444068917L;
    @Id
    @GeneratedValue(strategy = GenerationType.IDENTITY)
    @Column(name = "order_id")
    private Long id;

    @Column(name = "create_time")
    private LocalDateTime createTime;

    @Column(name = "payment")
    private BigDecimal payment;

    @ManyToOne(fetch = FetchType.LAZY)
    @JoinColumn(name = "user_id")
    private User user;

    /*一对多，级联保存*/
    @OneToMany(fetch = FetchType.LAZY, mappedBy = "order", cascade = CascadeType.PERSIST)
    private List<Sold> soldList;

}
```

售出记录类 Sold.java：

```java
@Accessors(chain = true)
@Setter
@Getter
@Entity
@Table(name = "t_sold")
public class Sold implements Serializable {

    private static final long serialVersionUID = 8713979233015105297L;

    @Id
    @GeneratedValue(strategy = GenerationType.IDENTITY)
    @Column(name = "sold_id")
    private Long id;

    @ManyToOne(fetch = FetchType.LAZY)
    @JoinColumn(name = "product_id")
    private Product product;

    @ManyToOne(fetch = FetchType.LAZY)
    @JoinColumn(name = "order_id")
    private Order order;

    @Column(name = "quantity")
    private Integer quantity;
```

}

产品类 Product.java：

```java
@Accessors(chain = true)
@Setter
@Getter
@Entity
@Table(name = "t_product")
public class Product implements Serializable {

    private static final long serialVersionUID = -4921932609253412600L;

    @Id
    @GeneratedValue(strategy = GenerationType.IDENTITY)
    @Column(name = "product_id")
    private Long id;

    @Column(name = "name", nullable = false, unique = true)
    @Length(min = 3, message = "*名称必须至少包含5个字符")
    private String name;

    @Column(name = "description")
    private String description;

    @Column(name = "quantity", nullable = false)
    @Min(value = 0, message = "*数量必须为非负数")
    private Integer quantity;

    @Column(name = "price", nullable = false)
    @DecimalMin(value = "0.00", message = "*价格必须为非负数")
    private BigDecimal price;

    @OneToMany(fetch = FetchType.LAZY, mappedBy = "product")
    private List<Sold> soldList;

}
```

9.4.3 用户注册

用户注册功能需要提供一个注册页面。注册页面需要填入用户名、密码、邮箱等用户信息。注册页面的模板文件 registration.html 如下：

```html
<!DOCTYPE HTML>
<html xmlns:th="http://www.thymeleaf.org">
<head>
    <div th:replace="/fragments/header :: header"/>
</head>
<body>
<div th:replace="/fragments/header :: navbar"/>
<div class="container">
    <!--如果注册成功，则弹窗提示-->
    <div class="alert alert-info" th:if="${successMessage}"
```

```html
th:utext="${successMessage}"></div>
        <div class="row" style="margin-top:20px">
           <div class="col-xs-12 col-sm-8 col-md-6 col-sm-offset-2 col-md-offset-3">
              <form autocomplete="off" action="#" th:action="@{/registration}"
                  th:object="${user}" method="post" role="form">
                  <div class="form-group">
                     <!--th:errors 结合 spring validation 使用。如果返回错误的字段为 name, 则显示该标签。下同-->
                     <label th:if="${#fields.hasErrors('name')}" th:errors="*{name}"
                           class="alert alert-danger"></label>
                     <input type="text" th:field="*{name}" placeholder="Name"
                           class="form-control input-lg"/>
                  </div>

                  <div class="form-group">
                     <label th:if="${#fields.hasErrors('lastName')}" th:errors="*{lastName}"
                           class="alert alert-danger"></label>
                     <input type="text" th:field="*{lastName}" placeholder="Last Name"
                           class="form-control input-lg"/>
                  </div>

                  <div class="form-group">
                     <label th:if="${#fields.hasErrors('email')}" th:errors="*{email}"
                           class="alert alert-danger"></label>
                     <input type="text" th:field="*{email}" placeholder="Email"
                           class="form-control input-lg"/>
                  </div>

                  <div class="form-group">
                     <label th:if="${#fields.hasErrors('password')}" th:errors="*{password}"
                           class="alert alert-danger"></label>
                     <input type="password" th:field="*{password}" placeholder="Password"
                           class="form-control input-lg"/>
                  </div>

                  <div class="form-group">
                     <label th:if="${#fields.hasErrors('username')}" th:errors="*{username}"
                           class="alert alert-danger"></label>
                     <input type="text" th:field="*{username}" placeholder="Username"
                           class="form-control input-lg"/>
                  </div>

                  <div class="row">
                     <div class="col-sm-3" style="float: none; margin: 0 auto;">
                        <input type="submit" class="btn btn-primary btn-block" value="Submit"/>
```

```
            </div>
        </div>
            </form>
    </div>
</div>

</div>
<div th:replace="/fragments/footer :: footer"/>
</body>
</html>
```

该页面引入了 header、navbar 和 footer 三个模板片段，这些部分将在章节末补齐。该页面中 th:object 获取对象，*{...}选择变量表达式用于从变量中获取对象。以上示例中的*{name}作用等同于${user.name}。

页面对应的控制器类 RegistrationController.java 如下：

```
@Controller
@RequiredArgsConstructor
public class RegistrationController {

    private final UserService userService;

    @GetMapping(value = "/registration")
    public ModelAndView registration() {
        ModelAndView modelAndView = new ModelAndView();
        User user = new User();
        modelAndView.addObject("user", user);
        modelAndView.setViewName("/registration");
        return modelAndView;
    }

    @PostMapping(value = "/registration")
    public ModelAndView createNewUser(@Valid User user, BindingResult bindingResult) {
        if (userService.findByEmail(user.getEmail()).isPresent()) {
            bindingResult.rejectValue("email", "error.user",
                    "已经有一个注册的用户使用了所提供的电子邮件");
        }
        if (userService.findByUsername(user.getUsername()).isPresent()) {
            bindingResult.rejectValue("username", "error.user",
                    "已经有一个注册的用户使用了所提供的用户名");
        }
        ModelAndView modelAndView = new ModelAndView();
        if (bindingResult.hasErrors()) {
            //注册失败，返回错误信息
            modelAndView.setViewName("/registration");
        } else {
            // 注册成功
            userService.saveUser(user);
            modelAndView.addObject("successMessage", "用户已成功注册");
            modelAndView.addObject("user", new User());
            modelAndView.setViewName("/registration");
        }
```

```
            return modelAndView;
        }
    }
```

控制器类分为两个方法，分别处理/registration 路径下的 Get 与 Post 请求。控制器在接收到 Get 请求后，返回一个带有 user 实例的视图。在接收到 Post 请求之后，使用 JPA 查询结合 User 实体类中标注的 Spring Validation 注解，对实体内容进行验证。验证的错误结果通过 bindingResult 返回。bindingResult.rejectValue 方法接收的三个参数，分别为错误字段、错误码以及提示信息。配合 Thymeleaf 的 th:errors 属性，可以完成对用户的错误提示。

处理用户信息的 UserService.java：

```
@Service
@RequiredArgsConstructor
public class UserService {

    private final UserRepository userRepository;
    private final RoleRepository roleRepository;
    private final PasswordEncoder passwordEncoder;

    private static final String USER_ROLE = "ROLE_USER";

    public Optional<User> findByUsername(String username) {
        return userRepository.findByUsername(username);
    }

    public Optional<User> findByEmail(String email) {
        return userRepository.findByEmail(email);
    }

    public User saveUser(User user) {
        user.setPassword(passwordEncoder.encode(user.getPassword()));
        user.setActive(1);
        user.setRoles(Collections.singletonList(roleRepository.findByRole(USER_ROLE)));
        return userRepository.saveAndFlush(user);
    }
}
```

实现好的注册页面如图 9.5 所示。

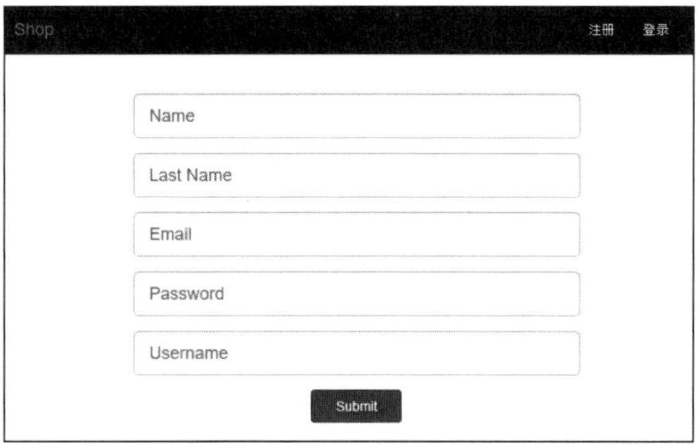

图 9.5　注册页面

9.4.4 用户登录

用户登录页面需要提供的内容有：供用户输入凭证的输入框以及信息提示。用户登录页面模板 login.html 如下：

```html
<!DOCTYPE html>
<html xmlns="http://www.w3.org/1999/xhtml"
xmlns:th="http://www.thymeleaf.org">

<head>
    <div th:replace="/fragments/header :: header"/>
</head>
<body>
<div th:replace="/fragments/header :: navbar"/>
<div class="container">

    <div class="row" style="margin-top:20px">
        <div class="col-xs-12 col-sm-8 col-md-6 col-sm-offset-2 col-md-offset-3">
            <form th:action="@{/login}" method="post">
                <fieldset>

                    <div th:if="${param.error}">
                        <div class="alert alert-danger">
                            无效的用户名或密码
                        </div>
                    </div>

                    <div th:if="${param.logout}">
                        <div class="alert alert-info">
                            已登出
                        </div>
                    </div>

                    <div class="form-group">
                        <input type="text" name="username" id="username" class="form-control input-lg"
                               placeholder="UserName" required="true" autofocus="true"/>
                    </div>

                    <div class="form-group">
                        <input type="password" name="password" id="password" class="form-control input-lg"
                               placeholder="Password" required="true"/>
                    </div>

                    <div class="row">
                        <div class="col-sm-3" style="float: none; margin: 0 auto;">
                            <input type="submit" class="btn btn-primary btn-block" value="Login"/>
                        </div>
```

```html
            </div>
        </fieldset>
      </form>
    </div>
  </div>

</div>

<div th:replace="/fragments/footer :: footer"/>

</body>
</html>
```

用户登录页面的控制器仅负责处理 Get 请求，即返回展示页。具体的验证功能交由 Spring Security 实现。登录控制器 LoginController.java：

```java
@Controller
public class LoginController {

    @GetMapping("/login")
    public String login(Principal principal) {
        if (principal != null) {
            return "redirect:/home";
        }
        return "/login";
    }

}
```

控制器将接收访客的权限，如果权限不为空，则代表访客为已登录的状态。因此使用 "redirect:/home" 将其重定向至主页。

要完成用户登录的功能，还需要配置 Spring Security。配置内容 SecurityConfig.java 如下：

```java
@Configuration
@RequiredArgsConstructor
public class SecurityConfig extends WebSecurityConfigurerAdapter {

    private final AccessDeniedHandler accessDeniedHandler;
    private final DataSource dataSource;
    @Value("${spring.queries.users-query}")
    private String usersQuery;
    @Value("${spring.queries.roles-query}")
    private String rolesQuery;

    @Override
    protected void configure(HttpSecurity http) throws Exception {
        http.csrf().disable()
                .authorizeRequests()
                .antMatchers("/", "/home", "/registration", "/error", "/h2-console/**"
                        , "/css/**", "/js/**", "/images/**", "/webjars/**").permitAll()
                .anyRequest().authenticated()
                .and()
```

```java
                .formLogin()
                    .loginPage("/login")
                    .defaultSuccessUrl("/home")
                    .permitAll()
                    .and()
                .logout()
                    .permitAll()
                    .and()
                .exceptionHandling().accessDeniedHandler(accessDeniedHandler)
                // 作用于 h2-console
                .and().headers().frameOptions().disable();
    }

    @Autowired
    public void configureGlobal(AuthenticationManagerBuilder auth) throws Exception {
        //用户信息使用数据库作为数据源
        auth.
                jdbcAuthentication()
                .usersByUsernameQuery(usersQuery)
                .authoritiesByUsernameQuery(rolesQuery)
                .dataSource(dataSource)
                .passwordEncoder(passwordEncoder());
    }

    /**
     * 默认返回 BCryptPasswordEncoder 的 PasswordEncoder
     */
    @Bean
    public PasswordEncoder passwordEncoder() {
        return new BCryptPasswordEncoder();
    }

}
```

其中 userQuery 与 rolesQuery 用于查询用户信息以及权限信息，对应的语句配置如下：

```
spring.queries.users-query=select username, password, active from t_user where username=?
spring.queries.roles-query=select u.username, r.role from t_user u inner join user_role ur on(u.user_id=ur.user_id) inner join t_role r on(ur.role_id=r.role_id) where u.username=?
```

登录失败的用户将会收到相关的错误提示。如果未处于登录成功的状态，直接访问受保护的链接，通常的做法是将请求重定向到错误页，错误页包含对于无效访问的提示信息。错误页 403.html 如下：

```html
<!DOCTYPE HTML>
<html xmlns:th="http://www.thymeleaf.org">

<head>
    <div th:replace="/fragments/header :: header"/>
</head>
```

```html
<body>
<div th:replace="/fragments/header :: navbar"/>

<div class="container">
    <div class="starter-template">
        <h1>403 - Access is denied</h1>
        <div th:inline="text">Hello '[[${#httpServletRequest.remoteUser}]]',
            没有访问该页面的权限
        </div>
    </div>
</div>

<div th:replace="/fragments/footer :: footer"/>

</body>
</html>
```

处理该类型请求的处理类 CustomAccessDeniedHandler.java 如下：

```java
@Slf4j
@Component
public class CustomAccessDeniedHandler implements AccessDeniedHandler {

    @Override
    public void handle(HttpServletRequest httpServletRequest,
HttpServletResponse httpServletResponse,
AccessDeniedException e) throws IOException {
        Authentication auth =
SecurityContextHolder.getContext().getAuthentication();
        if (auth != null) {
            log.info(String.format("User '%s' attempted to access the protected URL: %s", auth.getName(), httpServletRequest.getRequestURI()));
        }
        httpServletResponse.sendRedirect(httpServletRequest.getContextPath() + "/403");
    }
}
```

9.4.5 主页以及商品列表

在线商城的主页除了作为整个项目的用户入口之外，另外还内嵌了一个带有分页功能的商品列表。主页的页面模板 home.html 如下：

```html
<!DOCTYPE HTML>
<html xmlns:th="http://www.thymeleaf.org">

<head>
    <div th:replace="/fragments/header :: header"/>
</head>
<body>
<div th:replace="/fragments/header :: navbar"/>

<div class="container">
```

```html
        <div th:replace="/fragments/products :: products"/>
        <div th:replace="/fragments/pagination :: 
pagination(URLparameter='/home')"/>
    </div>

    <div th:replace="/fragments/footer :: footer"/>

</body>
</html>
```

除了页眉、导航条和页脚之外，主页还引入了用于展示商品列表的片段 products，以及分页组件 pagination。在引入 pagination 的同时，还向其传入 URLparameter 参数，用于 pagination 内部的功能实现。

商品列表片段 products.html 如下：

```html
<html xmlns="http://www.w3.org/1999/xhtml"
xmlns:th="http://www.thymeleaf.org"
xmlns:sec="http://www.thymeleaf.org/thymeleaf-extras-springsecurity5">

<head>
</head>
<body>

<div th:fragment="products">
    <div class="panel-default well" th:each="product : ${products}">
        <div class="panel-heading">
            <h1 th:text="${product.name}"></h1>
        </div>
        <h3 th:text="${product.description}" class="panel-body">Description</h3>
        <div class="row panel-footer">
            <div th:inline="text" class="col-md-2">价格：¥ [[${product.price}]]</div>
            <div th:inline="text" class="col-md-9">库存：[[${product.quantity}]]</div>
            <a th:href="@{'/shoppingCart/addProduct/{id}'(id=${product.id})}" class="col-md-1"
               sec:authorize="isAuthenticated()" th:if="${product.quantity}>0">
                <button type="button" class="btn btn-primary" th:text="购买">购买</button>
            </a>
        </div>
        <br/>
    </div>
</div>

</body>
</html>
```

分页组件 pagination.html 如下：

```html
<html xmlns="http://www.w3.org/1999/xhtml"
xmlns:th="http://www.thymeleaf.org">
```

```html
<head>
</head>
<body>
<div th:fragment="pagination">

    <!-- 翻页 -->
    <div class="pagination" th:with="baseUrl=${URLparameter}">
        <span th:if="${pager.indexOutOfBounds()}">
            页面超出范围。回到 <a class="pageLink" th:href="@{${baseUrl}(page=1)}">Home</a>.
        </span>

        <span th:unless="${pager.indexOutOfBounds()}">

            <span th:if="${pager.hasPrevious()}">
                <a class="pageLink" th:href="@{${baseUrl}(page=1)}">&laquo; first</a>
                <a class="pageLink" th:href="@{${baseUrl}(page=${pager.getPageIndex() - 1})}"> previous</a>
            </span>
            <span th:if="${pager.getTotalPages() != 1}"
                  th:text="'Page ' + ${pager.getPageIndex()} + ' of ' + ${pager.getTotalPages()} + '.'">
            </span>
            <span th:if="${pager.hasNext()}">
                <a class="pageLink" th:href="@{${baseUrl}(page=${pager.getPageIndex() + 1})}">next</a>
                <a class="pageLink" th:href="@{${baseUrl}(page=${pager.getTotalPages()})}">last &raquo;</a>
            </span>
        </span>
    </div>

</div>
</body>
</html>
```

使用分页组件需要向控制器传入当前页数，对应的展示内容将依赖于控制器返回的分页数据。主页的控制器 HomeController 如下：

```
@Controller
@RequiredArgsConstructor
public class HomeController {

    private final ProductRepository productRepository;

    @GetMapping(value = {"/home", "/"})
    public ModelAndView home(@RequestParam(value = "page", defaultValue = "1") Integer page) {
        //实际分页的参数与页面显示的页数不同，需要在页数的基础上减 1
        int evalPage = page - 1;
        //分页查询
        Page<Product> products = productRepository.findAll(PageRequest.of(evalPage, 5));
        //使用工具类从分页查询中提取分页组件所需的数据
```

```
            Pager pager = new Pager(products);
            ModelAndView modelAndView = new ModelAndView();
            modelAndView.addObject("products", products);
            modelAndView.addObject("pager", pager);
            modelAndView.setViewName("/home");
            return modelAndView;
        }
    }
```

分页过程中需要对分页结果进行判断，该步骤依赖于工具类 Pager。Pager.java 如下：

```
public class Pager {

    private final Page<Product> products;

    public Pager(Page<Product> products) {
        this.products = products;
    }
    /*获取当前页数*/
    public int getPageIndex() {
        return products.getNumber() + 1;
    }
    /*获取页大小*/
    public int getPageSize() {
        return products.getSize();
    }
    /*是否有下页*/
    public boolean hasNext() {
        return products.hasNext();
    }
    /*是否有上页*/
    public boolean hasPrevious() {
        return products.hasPrevious();
    }
    /*总页数*/
    public int getTotalPages() {
        return products.getTotalPages();
    }
    /*总元素个数*/
    public long getTotalElements() {
        return products.getTotalElements();
    }
    /*当前页是否超出页数上限*/
    public boolean indexOutOfBounds() {
        return this.getPageIndex() < 0 || this.getPageIndex() > this.getTotalElements();
    }
}
```

9.4.6　购物车

购物车功能需要提供购物车的页面以及结算相关的逻辑。购物车模板文件 shoppingCart.html 如下：

```html
<!DOCTYPE HTML>
<html xmlns:th="http://www.thymeleaf.org"
      xmlns:sec="http://www.thymeleaf.org/thymeleaf-extras-springsecurity4">

    <head>
        <div th:replace="/fragments/header :: header"/>
    </head>

    <body>

    <div th:replace="/fragments/header :: navbar"/>

    <div class="container">

        <h1 class="jumbotron">
            <span sec:authentication="name"></span>的购物车
        </h1>

    <div class="alert alert-info" th:if="${outOfStockMessage}"
     th:utext="${outOfStockMessage}"></div>

        <div class="panel-default well" th:each="product : ${products.entrySet()}">
            <div class="panel-heading">
                <h1><a th:text="${product.getKey().name}" th:href="@{'/product/' + ${product.getKey().id}}">Title</a></h1>
                <h3 th:text="${product.getKey().description}">介绍</h3>
            </div>
            <div class="row panel-body">
                <div th:inline="text" class="col-md-2">金额:¥ [[${product.getKey().price}]] </div>
                <div th:inline="text" class="col-md-9">数目: [[${product.getValue()}]]</div>
                <a th:href="@{'/shoppingCart/removeProduct/{id}'(id=${product.getKey().id})}" class="col-md-1">
                    <button type="button" class="btn btn-primary" th:text=""移除"">Remove</button>
                </a>
            </div>
            <br></br>
        </div>

        <div class="row panel-body">
            <h2 class="col-md-11" th:inline="text">总计: [[${total}]]</h2>
            <a th:href="@{'/shoppingCart/checkout'}" class="col-md-1">
                <button type="button" class="btn btn-danger" th:text=""付款"">Checkout</button>
            </a>
        </div>

    </div>

    <div th:replace="/fragments/footer :: footer"/>
```

```
    </body>
</html>
```

购物车控制器需要提供四个接口，分别用于展示、添加商品、移除商品以及结算。控制器 ShoppingCartController.java：

```
@Controller
@RequiredArgsConstructor
public class ShoppingCartController {

    private final ShoppingCartService shoppingCartService;
    private final UserRepository userRepository;
    private final ProductRepository productRepository;

    @GetMapping("/shoppingCart")
    public ModelAndView shoppingCart() {
        ModelAndView modelAndView = new ModelAndView("/shoppingCart");
        modelAndView.addObject("products", shoppingCartService.getProductsInCart());
        modelAndView.addObject("total", shoppingCartService.getTotal().toString());
        return modelAndView;
    }

    @GetMapping("/shoppingCart/addProduct/{productId}")
    public ModelAndView addProductToCart(@PathVariable("productId") Long productId) {
        productRepository.findById(productId).ifPresent(shoppingCartService::addProduct);
        return shoppingCart();
    }

    @GetMapping("/shoppingCart/removeProduct/{productId}")
    public ModelAndView removeProductFromCart(@PathVariable("productId") Long productId) {
        productRepository.findById(productId).ifPresent(shoppingCartService::removeProduct);
        return shoppingCart();
    }

    @GetMapping("/shoppingCart/checkout")
    public ModelAndView checkout(Principal principal) {
        Optional<User> optionalUser = userRepository.findByUsername(principal.getName());
        if (!optionalUser.isPresent()) {
            throw new HttpClientErrorException(HttpStatus.UNAUTHORIZED);
        }
        try {
            shoppingCartService.checkout(optionalUser.get());
        } catch (NotEnoughProductsInStockException e) {
            return shoppingCart().addObject("outOfStockMessage", e.getMessage());
        }
        return shoppingCart();
```

}
}

基于软件分层的设计，购物车控制器中主要的业务逻辑都交由 ShoppingCartService 完成。购物车服务类 ShoppingCartService.java 如下：

```java
@Service
@Scope(value = WebApplicationContext.SCOPE_SESSION, proxyMode =
ScopedProxyMode.TARGET_CLASS)
@Transactional
@RequiredArgsConstructor
public class ShoppingCartService {

    private final ProductRepository productRepository;
    private final OrderRepository orderRepository;

    private Map<Product, Integer> products = new HashMap<>();


    public void addProduct(Product product) {
        if (products.containsKey(product)) {
            products.replace(product, products.get(product) + 1);
        } else {
            products.put(product, 1);
        }
    }

    public void removeProduct(Product product) {
        if (products.containsKey(product)) {
            if (products.get(product) > 1)
                products.replace(product, products.get(product) - 1);
            else if (products.get(product) == 1) {
                products.remove(product);
            }
        }
    }

    public Map<Product, Integer> getProductsInCart() {
        return Collections.unmodifiableMap(products);
    }

    public void checkout(User user) throws NotEnoughProductsInStockException {
        ......
    }

    public BigDecimal getTotal() {
        return products.entrySet().stream()
                .map(entry -> entry.getKey().getPrice().
multiply(BigDecimal.valueOf(entry.getValue())))
                .reduce(BigDecimal::add)
                .orElse(BigDecimal.ZERO);
    }
}
```

该服务类由@Scope 更改了该服务类作为 Bean 的作用域。在 Spring 框架中，Bean 的作用域默认为单例。言下之意，在整个应用中该 Bean 的实例只会被创建一次。这种情况不利于每个用户对应购物车的状态维持，因为用户所添加的商品被加入 ShoppingCartService 的 products 列表属性中，ShoppingCartService 与用户会话需要保持一一对应的状态，单例模式 ShoppingCartService 并不能满足这种需求。

使用了@Scope(value = WebApplicationContext.SCOPE_SESSION, proxyMode = ScopedProxyMode.TARGET_CLASS) 之后，ShoppingCartService 的作用域将被切换至会话式，并通过 TARGET_CLASS 的方式（即 CGLIB）创建实例的代理。

ShoppingCartService 中结算相关逻辑实现如下：

```java
public void checkout(User user) throws NotEnoughProductsInStockException {
    Product product;
    Order order = new Order()
            .setCreateTime(LocalDateTime.now())
            .setUser(user);
    BigDecimal payment = BigDecimal.ZERO;
    List<Sold> soldList = new ArrayList<>();
    for (Map.Entry<Product, Integer> entry : products.entrySet()) {
        Product key = entry.getKey();
        Integer quantity = entry.getValue();
        Optional<Product> one = productRepository.findOne(Example.of(key));
        if (!one.isPresent()) {
            throw new IllegalArgumentException("");
        }
        product = one.get();
        if (product.getQuantity() < quantity) {
            throw new NotEnoughProductsInStockException(product);
        }
        entry.getKey().setQuantity(product.getQuantity() - quantity);
        soldList.add(new Sold()
                .setQuantity(quantity)
                .setProduct(key)
                .setOrder(order));
        payment = payment.add(key.getPrice());
    }
    order.setPayment(payment)
            .setSoldList(soldList);
    orderRepository.save(order);
    productRepository.saveAll(products.keySet());
    productRepository.flush();
    products.clear();
}
```

结算过程中将有可能抛出库存不足的异常，以起到提示用户的作用。库存不足异常类 NotEnoughProductsInStockException.java 如下：

```java
public class NotEnoughProductsInStockException extends Exception {

    private static final String DEFAULT_MESSAGE = "库存产品不足";

    public NotEnoughProductsInStockException() {
```

```
            super(DEFAULT_MESSAGE);
        }

        public NotEnoughProductsInStockException(Product product) {
            super(String.format("%s 库存不足。仅剩 %d 件。", product.getName(),
product.getQuantity()));
        }

}
```

9.4.7　页眉、导航条以及页脚

页眉与导航条模板片段 header.html 如下:

```
<html xmlns:th="http://www.thymeleaf.org"
      xmlns:sec="http://www.thymeleaf.org/thymeleaf-
extras-springsecurity5">
<head>
    <div th:fragment="header">
        <title  th:attr="data-custom=#{thymeleaf.app.title}">Shop</title>

        <link href="http://cdn.jsdelivr.net/webjars/bootstrap/
3.3.4/css/bootstrap.min.css"
              th:href="@{/webjars/bootstrap/3.3.7/css/bootstrap.min.css}"
              rel="stylesheet" media="screen"/>

        <script src="http://cdn.jsdelivr.net/webjars/jquery/
2.1.4/jquery.min.js"
                th:src="@{/webjars/jquery/2.1.4/jquery.min.js}"></script>
        <link rel="stylesheet" th:href="@{/css/main.css}"
              href="../../css/main.css"/>
    </div>
</head>

<body>
<div th:fragment="navbar">
    <nav class="navbar navbar-inverse">
        <div class="container">
            <div class="navbar-header">
                <a class="navbar-brand" th:href="@{/home}">Shop</a>
            </div>
            <div id="navbar" class="collapse navbar-collapse navbar-right">
                <!-- 仅在用户尚未通过身份验证时显示 shoppingCart -->
                <ul class="nav navbar-nav" sec:authorize="isAuthenticated()">
                    <li class="active"><a th:href="@{/shoppingCart}">购物车
</a></li>
                </ul>
                <!-- 仅在用户尚未通过身份验证时显示注册 -->
                <ul class="nav navbar-nav" sec:authorize="!isAuthenticated()">
                    <li class="active"><a th:href="@{/registration}">注册
</a></li>
                </ul>
                <!-- 仅在用户尚未通过身份验证时显示登录信息 -->
                <ul class="nav navbar-nav" sec:authorize="!isAuthenticated()">
```

```html
                <li class="active"><a th:href="@{/login}">登录</a></li>
            </ul>
            <!-- 仅在用户通过身份验证时显示注销 -->
            <ul class="nav navbar-nav" sec:authorize="isAuthenticated()">
                <li class="active"><a th:href="@{/logout}">登出</a></li>
            </ul>
        </div>
    </div>
</nav>
</div>

</body>
</html>
```

页脚模板片段 footer.html 如下：

```html
<html xmlns="http://www.w3.org/1999/xhtml" xmlns:th="http://www.thymeleaf.org"
      xmlns:sec="http://www.thymeleaf.org/thymeleaf-extras-springsecurity5">
<head>
</head>
<body>
<div th:fragment="footer">

    <hr/>
    <div class="container">
        <div class="row">
            <div class="col-sm-12 text-center">
                <span sec:authorize="isAuthenticated()">
                | Logged user: <span sec:authentication="name"></span> |
                Roles: <span sec:authentication="principal.authorities"></span> |
                    <a th:href="@{/logout}">Sign Out</a>
                </span>
            </div>
        </div>
    </div>
    <script type="text/javascript" src="webjars/bootstrap/3.3.7/js/bootstrap.min.js"></script>
</div>

</body>
</html>
```

9.5 测试与验证

在线商城的测试与验证环节分为集成测试与手工测试。集成测试关注于核心功能，手工测试验证整个业务流程。

9.5.1 测试数据

该程序测试所需的测试数据 import-h2.sql 如下：

```
-- 密码明文内容为"password"
INSERT INTO T_USER (user_id, password, email, username, name, last_name, active)
VALUES
    (1, '$2a$06$OAPObzhRdRXBCbk7Hj/ot.jY3zPwR8n7/mfLtKIgTzdJa4.6TwsIm',
'lester@mail.com', 'arnold', 'Cronin', 'Arnold',
    1);
-- 密码明文内容为"password"
INSERT INTO T_USER (user_id, password, email, username, name, last_name, active)
VALUES
    (2, '$2a$06$OAPObzhRdRXBCbk7Hj/ot.jY3zPwR8n7/mfLtKIgTzdJa4.6TwsIm',
'owen@gmail.com', 'rodney', 'Cecillia', 'Rodney', 1);
-- 密码明文内容为"password"
INSERT INTO T_USER (user_id, password, email, username, name, last_name, active)
VALUES (3, '$2a$06$OAPObzhRdRXBCbk7Hj/ot.jY3zPwR8n7/mfLtKIgTzdJa4.6TwsIm',
'carter@gmail.com', 'james ', 'Hosea', 'James', 1);

INSERT INTO T_ROLE (role_id, role)
VALUES (1, 'ROLE_ADMIN');
INSERT INTO T_ROLE (role_id, role)
VALUES (2, 'ROLE_USER');

INSERT INTO USER_ROLE (user_id, role_id)
VALUES (1, 1);
INSERT INTO USER_ROLE (user_id, role_id)
VALUES (1, 2);
INSERT INTO USER_ROLE (user_id, role_id)
VALUES (2, 2);
INSERT INTO USER_ROLE (user_id, role_id)
VALUES (3, 2);

INSERT INTO T_PRODUCT (name, description, quantity, price)
VALUES ('Java核心技术 卷Ⅰ 基础知识', 'Core Java 第11版', 1, 98);
INSERT INTO T_PRODUCT (name, description, quantity, price)
VALUES ('深入理解Java虚拟机：JVM高级特性与最佳实践（第3版）', '周志明虚拟机新作,第
3版新增内容近50%', 5, 89);
INSERT INTO T_PRODUCT (name, description, quantity, price)
VALUES ('Java编程思想（第4版）', 'Java学习必读经典,殿堂级著作！', 3, 70.20);
INSERT INTO T_PRODUCT (name, description, quantity, price)
VALUES ('Head First Java（中文版）', '10年畅销经典,累计印刷30多次。', 40, 54.50);
INSERT INTO T_PRODUCT (name, description, quantity, price)
VALUES ('Java开发手册（码出高效Java开发手册+阿里巴巴Java开发手册）', '引爆技术圈,
全球瞩目的中国计算机民族图书。', 80, 92.40);
INSERT INTO T_PRODUCT (name, description, quantity, price)
VALUES ('Java并发编程的艺术', '阿里技术专家/Java并发编程领域领军人物撰写。', 800,
38.80);
INSERT INTO T_PRODUCT (name, description, quantity, price)
VALUES ('Effective Java中文版（原书第3版）', 'Java之父鼎力推荐', 700, 87.30);
INSERT INTO T_PRODUCT (name, description, quantity, price)
VALUES ('Java项目开发实战入门（全彩版）', '一本让初学者通过项目实战开发学会编程的图书
```

```
', 500, 29.90);
    INSERT INTO T_PRODUCT (name, description, quantity, price)
    VALUES ('Java 精彩编程200例（全彩版）', '全彩图书，精选200个场景应用实例。', 1000, 39.90);
    INSERT INTO T_PRODUCT (name, description, quantity, price)
    VALUES ('Java 8 函数式编程', 'Java 编程思想从此向函数式编程转型', 10, 30.80);
```

9.5.2 集成测试

在线商城的核心功能在于购物车。对购物车功能的集成测试 CartTests.java 如下：

```
@SpringBootTest
@Transactional
class CartTests {

    private ProductRepository mockProductRepository;
    private MockMvc mockMvc;
    @Autowired
    private ShoppingCartService shoppingCartService;
    @Autowired
    private UserRepository userRepository;

    @BeforeEach
    void setup() {
        //如果视图名与路径名一致，需要做以下配置
        InternalResourceViewResolver viewResolver = new InternalResourceViewResolver();
        viewResolver.setPrefix("/templates");
        viewResolver.setSuffix(".html");
        //mock ProductRepository
        mockProductRepository = mock(ProductRepository.class);
        when(mockProductRepository.findById(anyLong()))
                .thenReturn(Optional.of(new Product()
                    .setId(1L)
                    .setDescription("该商品用于测试")
                    .setName("测试商品")
                    .setPrice(BigDecimal.TEN)
                    .setQuantity(1)
                ));
        mockMvc = MockMvcBuilders.standaloneSetup(
                new ShoppingCartController(shoppingCartService,
userRepository,
    mockProductRepository))
                .setViewResolvers(viewResolver)
                .build();
    }

    @Test
    @WithMockUser(username = "arnold")
    void testShoppingCart() throws Exception {
        //测试展示控制器
        ……
    }
```

```
@Test
@WithMockUser(username = "arnold")
void testAddProduct() throws Exception {
    //测试新增商品控制器
    ......
}

@Test
@WithMockUser(username = "arnold")
void testCheckout() throws Exception {
    //测试结算控制器
    ......
}
}
```

集成测试主要基于mockMvc对控制器各方法进行测试，以验证结果是否正常返回。各测试项的具体实现如下：

```
@Test
@WithMockUser(username = "arnold")
void testShoppingCart() throws Exception {
    //测试展示控制器
    mockMvc.perform(get("/shoppingCart"))
            .andExpect(status().isOk())
            .andExpect(view().name("/shoppingCart"))
            .andDo(MockMvcResultHandlers.print())
            .andReturn();
}

@Test
@WithMockUser(username = "arnold")
void testAddProduct() throws Exception {
    //测试新增商品控制器
    mockMvc.perform(get("/shoppingCart/addProduct/1"))
            .andExpect(status().isOk())
            .andExpect(view().name("/shoppingCart"))
            .andExpect(model().attribute("total", "10"))
            .andDo(MockMvcResultHandlers.print());
}

@Test
@WithMockUser(username = "arnold")
void testCheckout() throws Exception {
    //测试结算控制器
    mockMvc.perform(get("/shoppingCart/addProduct/1"))
            .andExpect(status().isOk())
            .andDo(MockMvcResultHandlers.print());
    mockMvc.perform(get("/shoppingCart/checkout"))
            .principal(new PrincipalImpl("arnold")))
            .andExpect(status().isOk())
            .andDo(MockMvcResultHandlers.print());
}
```

9.5.3 手工测试

手工测试的步骤按照一个普通用户访问程序的基本流程进行。

1. 商城主页

启动服务器并访问商城主页，默认地址 http://localhost:8080/，页面如图9.6所示。

图9.6 商城主页

单击商城主页底端的翻页，列表将展示下一页的内容。组件处将显示页数信息，如图9.7所示。

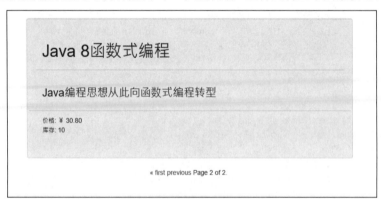

图9.7 翻页组件

2. 用户注册

单击导航栏中的注册按钮进入注册页面。直接单击Submit按钮，页面将弹出对每个输入项的提示，如图9.8所示。

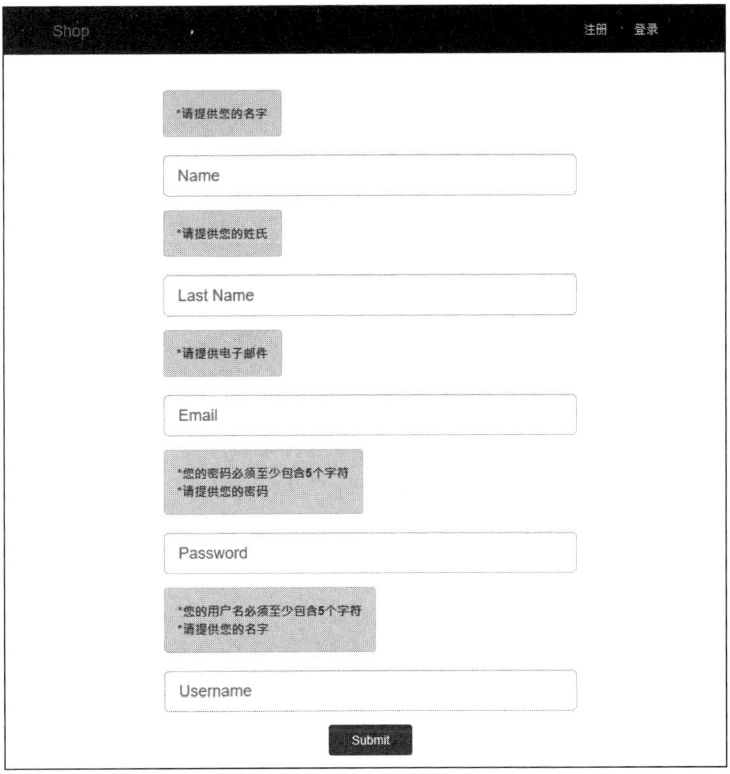

图 9.8　用户注册页面以及注册信息提示

按照页面提示填写注册信息并提交，将会弹出成功提示，如图 9.9 所示。

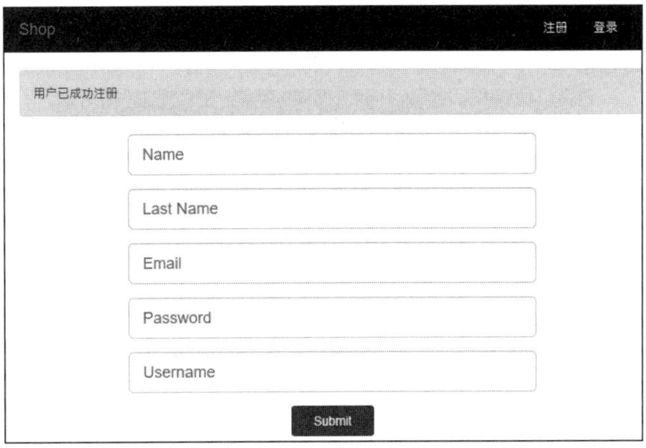

图 9.9　注册成功

3. 用户登录

单击导航栏中的登录键进入登录页面。在账号和密码输入框内填入不存在的账号和密码，页面中将提示登录无效，如图 9.10 所示。

填入正确的账号和密码登录后，页面将重定向至商城主页。登录状态下的商品列表将出现购买按钮，并且页脚会显示用户名、用户权限信息，如图 9.11 所示。

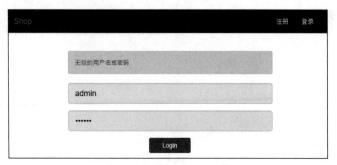

图 9.10　登录页面以及错误提示

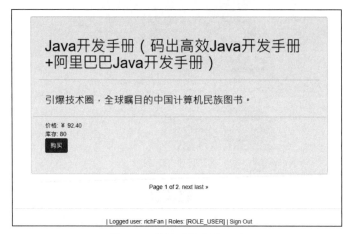

图 9.11　登录后的商城主页

4. 购物车

在商品列表单击购买按钮，页面将重定向至购物车/结算页面，如图 9.12 所示。

图 9.12　购物车

单击移除键可以移除目标商品，如图 9.13 所示。

图 9.13　移除商品后的购物车

选择超出库存范围的商品并付款，页面将返回相应提示，如图 9.14 所示。

图 9.14　库存不足提示

结算过后购买记录将写入数据库，商品列表中的库存将相应减少，如图 9.15 所示。

图 9.15 结算后的商品列表

第 10 章

实战 3：个人云盘

本章将基于 MinIO 的对象存储实现一款个人云盘。在实现过程中不仅讲解程序的设计与实现，还介绍 MinIO 的部署、基于 Spring Boot Starter 的第三方服务封装。

实战过程中所涉及的知识点如下：

- MinIO 的部署与使用
- 自定义 Spring Boot Starter
- 基于 Thymeleaf 的模板页面

10.1 架构设计

个人云盘程序依赖 MinIO 与 DBMS 共同实现数据的持久化，MinIO 负责用户文件的存储与管理，DBMS 负责业务数据的保存。访问层选择 Thymeleaf 开发模板页面，以便更好、更快地完成前端页面的编码。Spring Boot 部分使用 Web 模块、JPA、Spring Security 等框架以及依赖，以最终完成程序的功能实现。个人云盘的架构图如图 10.1 所示。

图 10.1 个人云盘架构图

10.2 框架搭建

框架搭建阶段，需要根据框架设计内容分析项目所需依赖。引入依赖项并且搭建项目所需的外部服务。在搭建之前，对存储层所依赖的外部服务——MinIO 有一个大概的了解，对功能的实现会非常有帮助。

10.2.1 MinIO 与对象存储

MinIO 是一个基于 Apache Licese v2.0 的开源对象存储服务。对象存储是以对象为数据单元的存储方式，对象可以是任何类型和任何大小的数据。对象存储中的所有对象都存储在单个平面地址空间中，而没有文件夹层次结构。

对象存储有以下优点：

- 易于访问：对象存储由元数据驱动。通过这种方式可以方便地对文件进行排序与搜索。
- 无限存储：对象存储空间不依赖于硬件。
- 降低成本：得益于对象存储的横向扩展性质，存储所有数据的成本更低廉。
- 资源优化：对象存储没有归档层次结构，并且元数据完全可自定义。与其他存储方式相比，对象存储的硬件限制要小得多。

MinIO 拥有以上所提及的优点，在标准硬件上，读/写速度上高达 183GB/s 和 171GB/s。MinIO 作为对象存储可以充当主存储层以处理各种复杂的工作负载。

MinIO 特有的功能特性如下：

- Amazon S3 兼容：MinIO 使用 Amazon S3 v2/v4 API。可以使用 MinIO SDK、MinIO Client、AWS SDK 和 AWS CLI 访问 Minio 服务器。
- 数据保护：MinIO 使用 MinIO Erasure Code 来防止硬件故障。也许会损坏一半以上的 Driver，但是仍然可以从中恢复数据。
- 高度可用：MinIO 服务器可以容忍分布式设置中高达（N/2）-1 节点的故障。而且可以配置 MinIO 服务器在 MinIO 与任意 Amazon S3 兼容服务器之间存储数据。
- Lambda 计算：MinIO 服务器通过其兼容 AWS SNS / SQS 的事件通知服务触发 Lambda 功能。支持的目标是消息队列，如 Kafka、NATS、AMQP、MQTT、Webhooks 以及 ElasticSearch、Redis、PostgreSQL 和 MySQL 等数据库。
- 加密和防篡改：MinIO 为加密数据提供了机密性、完整性和真实性保证，而且性能开销微乎其微。使用 AES-256-GCM、ChaCha20-Poly1305 和 AES-CBC 支持服务端和客户端加密。加密的对象使用 AEAD 服务端加密进行防篡改。
- 可对接后端存储：除了 MinIO 自己的文件系统，还支持 DAS、JBODs、NAS、Google 云存储和 Azure Blob 存储。
- SDK 支持：基于 MinIO 轻量的特点，它得到类似 Java、Python 或 Go 等语言的 SDK 支持。

10.2.2 MinIO 部署与使用

MinIO 有几种不同的部署方式，本实战过程中推荐的部署方式为 Docker 部署。参考命令如下：

```
docker pull minio/minio
docker run -p 9000:9000 \
 -e "MINIO_ACCESS_KEY=AKIAIOSFODNN7EXAMPLE" \
 -e "MINIO_SECRET_KEY=wJalrXUtnFEMI/K7MDENG/bPxRfiCYEXAMPLEKEY" \
 minio/minio server /data
```

以上命令中的参数 MINIO_ACCESS_KEY 与 MINIO_SECRET_KEY，分别为用于登录 MinIO 管理页面的用户名与密码。

输入以上命令即可完成 MinIO 的部署，访问 http://127.0.0.1:9000 即可访问 MinIO 的管理页面，如图 10.2 所示。

图 10.2　MinIO 管理页面

使用 MinIO 将涉及的一些对象存储概念，说明如下：

- Bucket：存储桶，对应文件系统的目录。
- Object：对象，对应文件系统中的文件。
- Endpoint：对象存储的访问链接。
- Region：数据区域。
- AccessKey：访问秘钥。

在 MinIO 管理页面可以对 MinIO 的存储内容进行基础维护，包括创建存储桶、文件的上传下载等。

10.2.3　项目依赖项与软件包结构

根据需求选择所需的依赖项。展示层需要引入 Thymeleaf 相关依赖，存储层引入 JPA 以及 MinIO

相关依赖。项目内 pom.xml 中主要依赖项如下：

```xml
<!--Web-->
<dependency>
    <groupId>org.springframework.boot</groupId>
    <artifactId>spring-boot-starter-web</artifactId>
</dependency>
<!--JPA-->
<dependency>
    <groupId>org.springframework.boot</groupId>
    <artifactId>spring-boot-starter-data-jpa</artifactId>
</dependency>
<!--验证相关依赖-->
<dependency>
    <groupId>org.springframework.boot</groupId>
    <artifactId>spring-boot-starter-validation</artifactId>
</dependency>
<!--安全框架-->
<dependency>
    <groupId>org.springframework.boot</groupId>
    <artifactId>spring-boot-starter-security</artifactId>
</dependency>
<!--H2 数据库-->
<dependency>
    <groupId>com.h2database</groupId>
    <artifactId>h2</artifactId>
    <scope>runtime</scope>
</dependency>
<dependency>
    <groupId>org.springframework.boot</groupId>
    <artifactId>spring-boot-starter-thymeleaf</artifactId>
</dependency>
<dependency>
    <groupId>org.thymeleaf.extras</groupId>
    <artifactId>thymeleaf-extras-springsecurity5</artifactId>
</dependency>
<!--可选项 使用Maven 管理前端资源-->
<dependency>
    <groupId>org.webjars</groupId>
    <artifactId>bootstrap</artifactId>
    <version>4.1.3</version>
</dependency>
<dependency>
    <groupId>org.webjars</groupId>
    <artifactId>jquery</artifactId>
    <version>3.3.1</version>
</dependency>
<dependency>
    <groupId>org.webjars.npm</groupId>
    <artifactId>popper.js</artifactId>
    <version>1.16.1-lts</version>
</dependency>
```

个人云盘的软件包结构如图 10.3 所示。

```
src
├─main
│  ├─java
│  │  └─com
│  │     └─example
│  │        └─storage
│  │           │  CloudStorageApplication.java
│  │           │
│  │           ├─controller
│  │           │
│  │           ├─model
│  │           │
│  │           ├─repository
│  │           │
│  │           ├─config
│  │           │
│  │           └─service
│  │
│  └─resources
│     │
│     ├─static
│     │  └─js
│     │
│     └─templates
│
└─test
```

图 10.3　个人云盘的软件包结构

包结构说明：

- config：包含项目的配置文件。
- controller：包含控制器文件。
- model：包含模型对象。
- repository：包含数据访问层文件。
- service：包含服务类。
- resources：包含静态文件、模板以及项目初始化所需的 SQL 语句。

10.3　数据库设计

个人云盘所需存储的信息分别有用户信息、文件信息以及笔记信息。每个信息对应一张关系型数据库的表。数据库 E-R 图如图 10.4 所示。

表名以 "t_" 作前缀，表说明如下：

- t_user：用户信息表，用于存储已注册用户的相关信息。
- t_file：文件信息表，用于存储用户上传至云盘内的文件信息，其中包括文件名、存储对象名、访问地址以及文件大小。
- t_note：笔记信息表，用于存储用户保存至云盘的笔记信息。

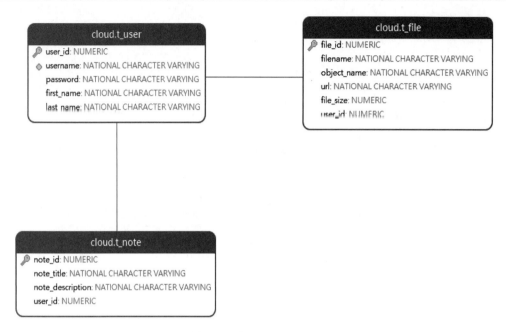

图 10.4 个人云盘 E-R 图

10.4 功能实现

该程序的核心功能在于文件存储的相关操作，这一部分主要依赖于 MinIO 实现。MinIO 官方提供了一套功能齐全的 Java SDK，该 SDK 覆盖了对 MinIO 的所有操作，非常强大并且灵活。但在使用过程中不免会有不少样板代码。一方面为了消除样板代码，在除了该项目之外也能让 MinIO 开箱即用；另一方面为了演示如何自定义一个 Spring Boot Starter，在了解过 MinIO 的 SDK 之后将会把它以 Starter 的形式进行封装。

10.4.1 MinIO Java SDK 简介

MinIO 提供了一套兼容 Amazon S3 标准的 Java 客户端 SDK，对应的依赖如下：

```
<dependency>
    <groupId>io.minio</groupId>
    <artifactId>minio</artifactId>
    <version>8.0.0</version>
</dependency>
```

使用该 SDK 需要提供三个参数才能连接到 MinIO 服务，参数说明：

- Endpoint：对象存储服务的 URL。
- Access Key：Access Key 相当于用户名，作为存储账户的唯一标识。
- Secret Key：账号密码。

MinIO 团队为了帮助开发人员掌握 MinIO SDK 相关开发与测试,在 https://play.min.io 部署了一个免费的 MinIO 服务。该 URL 可以作为示例中的 EndPoint 使用。一个基础的文件上传示例如下:

```java
import io.minio.BucketExistsArgs;
import io.minio.MakeBucketArgs;
import io.minio.MinioClient;
import io.minio.UploadObjectArgs;
import io.minio.errors.MinioException;
import java.io.IOException;
import java.security.InvalidKeyException;
import java.security.NoSuchAlgorithmException;

public class FileUploader {
  public static void main(String[] args)
      throws IOException, NoSuchAlgorithmException, InvalidKeyException {
    try {
      // 使用以下参数连接 MinIO 提供的公共服务
      MinioClient minioClient =
          MinioClient.builder()
              .endpoint("https://play.min.io")
              .credentials("Q3AM3UQ867SPQQA43P2F",
"zuf+tfteSlswRu7BJ86wekitnifILbZam1KYY3TG")
              .build();

      // 如果'asiatrip' 存储桶不存在,则创建它
      boolean found =
          minioClient.bucketExists(BucketExistsArgs.builder().
bucket("asiatrip").build());
      if (!found) {
        // 创建一个名为'asiatrip'的存储桶
        minioClient.makeBucket(MakeBucketArgs.builder().
bucket("asiatrip").build());
      } else {
        System.out.println("Bucket 'asiatrip' already exists.");
      }

      // 将文件'/home/user/Photos/asiaphotos.zip' 以'asiaphotos-2015.zip' 为对象名上传至桶
      // 'asiatrip'.
      minioClient.uploadObject(
          UploadObjectArgs.builder()
              .bucket("asiatrip")
              .object("asiaphotos.zip")
              .filename("/home/user/Photos/asiaphotos.zip")
              .build());
      System.out.println(
          "'/home/user/Photos/asiaphotos.zip' is successfully uploaded as "
              + "object 'asiaphotos-2015.zip' to bucket 'asiatrip'.");
    } catch (MinioException e) {
      System.out.println("Error occurred: " + e);
    }
  }
}
```

以上示例演示了连接 MinIO 服务、创建存储桶以及通过指定文件名的形式上传文件。

> **注 意**
>
> 最新版本通过建造者模式构造所需的参数。

为了顺利实现一个云盘的基本功能，另外需要借助 MinIO 实现的功能还有：通过文件流形式上传文件、通过文件流形式下载文件以及删除文件。

（1）上传文件

```java
// 以流的形式上传已知大小的文件
   minioClient.putObject(PutObjectArgs.builder().bucket("my-bucketname").
object("my-objectname").stream(inputStream, size, -1)
        .contentType("video/mp4")
        .build());

// 以流的形式上传未知大小的文件
   minioClient.putObject(PutObjectArgs.builder().bucket("my-bucketname").
object("my-objectname").stream(inputStream, -1, 10485760)
        .contentType("video/mp4")
        .build());

// 创建以'/'结尾的对象
   minioClient.putObject(
      PutObjectArgs.builder().bucket("my-bucketname").
object("path/to/").stream(new ByteArrayInputStream(new byte[] {}), 0, -1)
        .build());

// 以流的形式上传文件、头信息和用户元数据
   Map<String, String> headers = new HashMap<>();
   headers.put("X-Amz-Storage-Class", "REDUCED_REDUNDANCY");
   Map<String, String> userMetadata = new HashMap<>();
   userMetadata.put("My-Project", "Project One");
   minioClient.putObject(PutObjectArgs.builder().bucket("my-bucketname").
object("my-objectname").stream(inputStream, size, -1)
        .headers(headers)
        .userMetadata(userMetadata)
        .build());

// 以流的形式上传服务端加密文件
   minioClient.putObject(
      PutObjectArgs.builder().bucket("my-bucketname").
object("my-objectname").stream(inputStream, size, -1)
        .sse(sse)
        .build());
```

（2）下载文件

```java
// 通过给出的存储桶名与对象名获取输入流
   try (InputStream stream = minioClient.getObject(
     GetObjectArgs.builder()
     .bucket("my-bucketname")
     .object("my-objectname")
     .build()) {
     // Read data from stream
   }
```

```
// 根据偏转量获取对象
try (InputStream stream = minioClient.getObject(
  GetObjectArgs.builder()
    .bucket("my-bucketname")
    .object("my-objectname")
    .offset(1024L)
    .build()) {
  // Read data from stream
}

// 根据偏转量和长度获取对象
try (InputStream stream = minioClient.getObject(
  GetObjectArgs.builder()
    .bucket("my-bucketname")
    .object("my-objectname")
    .offset(1024L)
    .length(4096L)
    .build()) {
  // Read data from stream
}

// 获取一个服务端加密的对象
try (InputStream stream = minioClient.getObject(
  GetObjectArgs.builder()
    .bucket("my-bucketname")
    .object("my-objectname")
    .ssec(ssec)
    .build()) {
  // Read data from stream
}

// 根据偏转量和长度获取一个服务端加密的对象
try (InputStream stream = minioClient.getObject(
  GetObjectArgs.builder()
    .bucket("my-bucketname")
    .object("my-objectname")
    .offset(1024L)
    .length(4096L)
    .ssec(ssec)
    .build()) {
  // Read data from stream
}
```

（3）删除文件

```
try {
    // 从 mybucket 中删除 myobject
    minioClient.removeObject("mybucket", "myobject");
    System.out.println("successfully removed mybucket/myobject");
} catch (MinioException e) {
    System.out.println("Error: " + e);
}
```

10.4.2 实现 MinIO Starter

为了获得一个在其他项目中也能方便使用的 MinIO 客户端功能，本小节将基于 MinIO Java SDK 封装一个自定义的 Spring Boot Starter。Starter 是一个独立的模块，需要另外新建项目进行编写。因为该 Starter 基于 MinIO Java SDK 封装的，所以根据社区的命名规律将其命名为 minio-spring-boot-starter。具体编写步骤说明如下。

（1）声明依赖

minio-spring-boot-starter 的 pom.xml 如下：

```xml
<?xml version="1.0" encoding="UTF-8"?>
<project xmlns="http://maven.apache.org/POM/4.0.0"
         xmlns:xsi="http://www.w3.org/2001/XMLSchema-instance"
         xsi:schemaLocation="http://maven.apache.org/POM/4.0.0 http://maven.apache.org/xsd/maven-4.0.0.xsd">
    <modelVersion>4.0.0</modelVersion>
    <parent>
        <groupId>org.springframework.boot</groupId>
        <artifactId>spring-boot-starters</artifactId>
        <version>2.2.0.RELEASE</version>
    </parent>

    <groupId>org.example</groupId>
    <artifactId>minio-spring-boot-starter</artifactId>
    <version>1.0-SNAPSHOT</version>
    <properties>
        <minio.version>8.0.0</minio.version>
    </properties>

    <dependencies>
        <dependency>
            <groupId>org.springframework.boot</groupId>
            <artifactId>spring-boot-starter</artifactId>
        </dependency>
        <dependency>
            <groupId>io.minio</groupId>
            <artifactId>minio</artifactId>
            <version>${minio.version}</version>
        </dependency>
        <dependency>
            <groupId>org.springframework.boot</groupId>
            <artifactId>spring-boot-starter-test</artifactId>
            <scope>test</scope>
            <exclusions>
                <exclusion>
                    <groupId>org.junit.vintage</groupId>
                    <artifactId>junit-vintage-engine</artifactId>
                </exclusion>
            </exclusions>
        </dependency>
    </dependencies>
</project>
```

（2）编写配置实体类

minio-spring-boot-starter 的目的在于减少样板代码，因此将会把灵活变动的部分以配置的形式提取出来。配置相关的实体类 MinioConfigurationProperties.java 如下：

```java
@ConfigurationProperties("spring.minio")
public class MinioConfigurationProperties {

    private String endpoint;
    private String accessKey;
    private String secretKey;
    private String bucket;

    //……省略 getter、setter
}
```

@ConfigurationProperties 将会把配置信息中带有 spring.minio 前缀的内容，映射成配置实体的实例，以便系统内对配置信息的调用。

（3）编写配置类

调用 MinIO 服务需要用到 MinioClient 的对象实例，因此可以将其在配置中提供对应的 JavaBean。配置类 MinioConfiguration.java 如下：

```java
@Configuration
@ConditionalOnClass(MinioClient.class)
@EnableConfigurationProperties(MinioConfigurationProperties.class)
@ComponentScan("com.example.minio")
public class MinioConfiguration {

    @Autowired
    private MinioConfigurationProperties minioConfigurationProperties;

    @Bean
    public MinioClient minioClient() {
        return MinioClient.builder()
                .endpoint(minioConfigurationProperties.getEndpoint())
                .credentials(minioConfigurationProperties.getAccessKey(), minioConfigurationProperties.getSecretKey())
                .build();
    }
}
```

其中@ConditionalOnClass 作用是声明，如果 classpath 中有 MinioClient 类，才会加载 MinioConfiguration。@EnableConfigurationProperties 的作用是启用配置实体 MinioConfigurationProperties。

（4）编写服务类

服务类 MinioService.java 如下：

```java
@Service
public class MinioService implements ApplicationRunner {

    private static final Logger LOGGER =
```

```
LoggerFactory.getLogger(MinioConfiguration.class);

    @Autowired
    private MinioClient minioClient;

    @Autowired
    private MinioConfigurationProperties minioConfigurationProperties;

    /**
     * 程序启动时自动创建 bucket
     */
    public void run(ApplicationArguments args) throws Exception {
        if (!minioClient.bucketExists(BucketExistsArgs
                .builder()
                .bucket(minioConfigurationProperties.getBucket())
                .build())) {
            minioClient.makeBucket(MakeBucketArgs
                    .builder()
                    .bucket(minioConfigurationProperties.getBucket())
                    .build());
            LOGGER.info("{} is created successfully",
minioConfigurationProperties.getBucket());
        }
    }

    public ObjectWriteResponse uploadObject(String objectName, InputStream
inputStream) throws Exception {
        //以输入流形式上传文件
    }

    public void removeObject(String objectName) throws Exception {
        //移除对象（文件）
    }

    public InputStream getObject(String objectName) throws IOException,
InvalidKeyException, InvalidResponseException,InsufficientDataException,
NoSuchAlgorithmException, ServerException, InternalException, XmlParserException,
ErrorResponseException {
        //获取文件，下载文件
    }

    public void removeBucket() throws Exception {
        //移除存储桶
    }

}
```

服务类实现了 **ApplicationRunner** 接口，目的是在程序启动时自动检测并创建一个存储桶。另外还封装了几个 MinIO Java SDK 的接口。

以输入流形式上传文件：

```
    public ObjectWriteResponse uploadObject(String objectName, InputStream
inputStream) throws Exception {
        //以输入流形式上传文件
        return minioClient.putObject(
```

```java
                PutObjectArgs.builder()
                        .bucket(minioConfigurationProperties.getBucket())
                        .object(objectName)
                        .stream(inputStream, inputStream.available(), -1)
                        .build());
}
```

删除文件：

```java
public void removeObject(String objectName) throws Exception {
    //移除对象（文件）
    minioClient.removeObject(
            RemoveObjectArgs.builder()
                    .bucket(minioConfigurationProperties.getBucket())
                    .object(objectName)
                    .build());
}
```

下载文件：

```java
public InputStream getObject(String objectName) throws IOException,
InvalidKeyException, InvalidResponseException,InsufficientDataException,
NoSuchAlgorithmException, ServerException, InternalException, XmlParserException,
ErrorResponseException {
    //获取文件，下载文件
    minioClient.statObject(StatObjectArgs.builder()
            .bucket(minioConfigurationProperties.getBucket())
            .object(objectName)
            .build());
    return minioClient.getObject(
            GetObjectArgs.builder()
                    .bucket(minioConfigurationProperties.getBucket())
                    .object(objectName)
                    .build());
}
```

移除存储桶：

```java
public void removeBucket() throws Exception {
    //移除存储桶
    boolean found = minioClient.bucketExists(BucketExistsArgs
            .builder()
            .bucket(minioConfigurationProperties.getBucket())
            .build());
    if (found) {
        minioClient.removeBucket(RemoveBucketArgs
                .builder()
                .bucket(minioConfigurationProperties.getBucket())
                .build());
        LOGGER.info("{} removed successfully",
minioConfigurationProperties.getBucket());
    } else {
        LOGGER.info("{} does not exist",
minioConfigurationProperties.getBucket());
    }
}
```

（5）编写 Spring Factories

Spring Factories 类似于 Java SPI，是 Spring Boot 中的一种扩展机制。编写该文件用于指定自动配置类，文件路径为 resource/META-INF/spring.factories。内容如下：

```
org.springframework.boot.autoconfigure.EnableAutoConfiguration=com.example
.minio.MinioConfiguration
```

（6）发布 minio-spring-boot-starter

spring.factories 配置完成之后，意味着该 starter 可以发布了。发布软件包的形式有很多，本案例中仅需要使用"mvn clean install"命令，将该软件包安装至本地 Maven 仓库，即可开放给其他项目使用。

10.4.3 实体类

基于数据库设计，编写程序所需的实体类。用户类 TUser.java：

```java
@Accessors(chain = true)
@Setter
@Getter
@Entity
@Table(name = "t_user")
public class TUser implements Serializable {

    private static final long serialVersionUID = 2749493794886279420L;
    @Id
    @GeneratedValue(strategy = GenerationType.IDENTITY)
    @Column(name = "user_id")
    private Long id;

    @Column(name = "password", nullable = false)
    @Length(min = 5, message = "*您的密码必须至少包含5个字符")
    @NotEmpty(message = "*请提供您的密码")
    @JsonIgnore
    private String password;

    @Column(name = "username", nullable = false, unique = true)
    @Length(min = 5, message = "*您的用户名必须至少包含5个字符")
    @NotEmpty(message = "*请提供您的用户名")
    private String username;

    @Column(name = "first_name")
    @NotEmpty(message = "*请提供您的名字")
    private String firstName;

    @Column(name = "last_name")
    @NotEmpty(message = "*请提供您的姓氏")
    private String lastName;

    @OneToMany(fetch = FetchType.LAZY,mappedBy = "user")
    private List<TNote> noteList;

    @OneToMany(fetch = FetchType.LAZY,mappedBy = "user")
```

```
    private List<TFile> fileList;
}
```

用户类与笔记以及文件均是一对多的关系。笔记类 TNote.java：

```
@Accessors(chain = true)
@Setter
@Getter
@Entity
@Table(name = "t_note")
public class TNote implements Serializable {

private static final long serialVersionUID = 3706652240480176782L;

    @Id
    @GeneratedValue(strategy = GenerationType.IDENTITY)
    @Column(name = "note_id")
    private Long id;

    @Column(name = "note_title", nullable = false)
    @NotEmpty(message = "*请提供笔记标题")
    @JsonIgnore
    private String noteTitle;

    @Column(name = "note_description")
    @NotEmpty(message = "*请提供笔记描述")
    @JsonIgnore
    private String noteDescription;

    @ManyToOne(fetch = FetchType.LAZY)
    @JoinColumn(name = "user_id")
    @JsonIgnore
    private TUser user;
}
```

文件类 TFile.java：

```
@Accessors(chain = true)
@Setter
@Getter
@Entity
@Table(name = "t_file")
public class TFile implements Serializable {

    private static final long serialVersionUID = -6499207403641470328L;
    @Id
    @GeneratedValue(strategy = GenerationType.IDENTITY)
    @Column(name = "file_id")
    private Long id;

    @Column(name = "filename", nullable = false)
    @NotEmpty(message = "*请提供文件名")
    @JsonIgnore
    private String filename;
```

```java
    @Column(name = "object_name", nullable = false)
    private String objectName;

    @Column(name = "url", nullable = false, unique = true)
    private String url;

    @Column(name = "file_size")
    private Long fileSize;

    @ManyToOne(fetch = FetchType.LAZY)
    @JoinColumn(name = "user_id")
    @JsonIgnore
    private TUser user;
}
```

10.4.4 用户注册

提供用户注册功能的页面,需要对用户提供输入姓名以及账号和密码的输入框,控制器则对用户提交的这些信息进行处理。注册页面示例 signup.html 如下:

```html
<!DOCTYPE html>
<html lang="en" xmlns="http://www.w3.org/1999/xhtml"
      xmlns:th="https://www.thymeleaf.org">
<head>
    <meta charset="utf-8">
    <meta name="viewport" content="width=device-width, initial-scale=1, shrink-to-fit=no">
    <link href="http://cdn.jsdelivr.net/webjars/bootstrap/4.1.3/css/bootstrap.min.css"
          th:href="@{/webjars/bootstrap/4.1.3/css/bootstrap.min.css}"
          rel="stylesheet" media="screen"/>
    <title>注册</title>
</head>
<body class="p-3 mb-2 bg-light text-black">
    <div class="container justify-content-center w-50 p-3" style="background-color: #eeeeee; margin-top: 5em;">
        <div class="form-group">
            <label><a th:href="@{/login}">返回登录</a></label>
        </div>
        <h1 class="display-5">注册</h1>
        <form th:action="@{/register}" method="POST">
            <div th:if="${param.success}" class="alert alert-dark">
                您已成功注册。请返回 <a th:href="@{/login}">登录</a> 页面
            </div>
            <div th:if="${param.error}" class="alert alert-danger">
                <span>错误</span>
            </div>
            <div class="form-row">
                <div class="form-group col-md-6">
                    <label for="inputFirstName">名</label>
                    <input type="input" name="firstName" class="form-control" id="inputFirstName" placeholder="Enter First Name" maxlength="20" required>
```

```html
            </div>
            <div class="form-group col-md-6">
                <label for="inputLastName">姓</label>
                <input type="input" name="lastName" class="form-control" id="inputLastName" placeholder="Enter Last Name" maxlength="20" required>
            </div>
        </div>
        <div class="form-row">
            <div class="form-group col-md-6">
                <label for="inputUsername">用户名</label>
                <input type="input" name="username" class="form-control" id="inputUsername" placeholder="Enter Username" maxlength="20" required>
            </div>
            <div class="form-group col-md-6">
                <label for="inputPassword">密码</label>
                <input type="password" name="password" class="form-control" id="inputPassword" placeholder="Enter Password" maxlength="20" required>
            </div>
        </div>
        <button id = "signup" type="submit" class="btn btn-primary">注册</button>
    </form>
  </div>
 </body>
</html>
```

页面的静态资源使用 webjar 引入，在其他页面中也是相同的处理方式。页面将接收控制器返回的 success 或者 error 参数，根据这些参数显示成功或失败的提示。注册页面对应的控制器 UserController.java 如下：

```java
@Controller
@RequiredArgsConstructor
public class UserController {

    private final UserService userService;

    @PostMapping("/register")
    public String register(@ModelAttribute("user") TUser user) {
        if (user == null) {
            return "redirect:signup";
        }
        try {
            userService.register(user);
        } catch (Exception e) {
            e.printStackTrace();
            return "redirect:signup?error";
        }
        return "redirect:signup?success";
    }
}
```

控制器调用 UserService 提供的注册方法，处理接收到的用户信息。User 相关服务类 UserService.java 如下：

```
@Service
@RequiredArgsConstructor
public class UserService {

    private final PasswordEncoder passwordEncoder;
    private final UserRepository userRepository;

    public TUser register(TUser user) throws Exception {
        String encodedPassword = passwordEncoder.encode(user.getPassword());
        user.setPassword(encodedPassword);
        userRepository.save(user);
        return user;
    }
}
```

其中 PasswordEncoder 将在 Spring Security 相关的配置类中提供对应的 JavaBean。

10.4.5 用户登录

用户登录页面需要提供的有：供用户输入凭证的输入框以及信息提示。页面示例文件 login.html 的源代码如下：

```
<!DOCTYPE html>
<html lang="en" xmlns="http://www.w3.org/1999/xhtml"
xmlns:th="https://www.thymeleaf.org">
    <head>
        <meta charset="utf-8">
        <meta name="viewport" content="width=device-width, initial-scale=1, shrink-to-fit=no">
        <link href="http://cdn.jsdelivr.net/webjars/bootstrap/4.1.3/css/bootstrap.min.css"
              th:href="@{/webjars/bootstrap/4.1.3/css/bootstrap.min.css}"
              rel="stylesheet" media="screen"/>
        <title>用户登录</title>
    </head>
    <body class="p-3 mb-2 bg-light text-black">
    <div class="container justify-content-center w-25 p-3" style="background-color: #eeeeee; margin-top: 5em;">
        <h1 class="display-5">登录</h1>
        <form th:action="@{/login}" method="POST">
            <div th:if="${param.error}" class="alert alert-danger">
                无效的用户名或密码
            </div>
            <div th:if="${param.logout}" class="alert alert-dark">
                您已登出
            </div>
            <div class="form-group">
                <label for="inputUsername">用户名</label>
                <input type="input" class="form-control" name="username"
id="inputUsername" placeholder="Enter Username"
                       maxlength="20" required>
            </div>
            <div class="form-group">
```

```html
            <label for="inputPassword">密码</label>
            <input type="password" class="form-control" name="password" id="inputPassword" placeholder="Enter Password"
                maxlength="20" required>
        </div>
        <button id="login" type="submit" class="btn btn-primary">登录</button>
    </form>

    <div class="form-group" style="margin-top: 0.5em;">
        <label><a th:href="@{/signup}">单击注册</a></label>
    </div>

</div>
</body>
</html>
```

用户配置的功能通过配置 Spring Sercurity 实现，Spring Sercurity 配置类 SecurityConfiguration.java 如下：

```java
@Configuration
@EnableWebSecurity
@RequiredArgsConstructor
public class SecurityConfiguration extends WebSecurityConfigurerAdapter {

    private final DataSource dataSource;
    private final UserDetailsService userDetailsService;

    @Override
    protected void configure(HttpSecurity http) throws Exception {
        http.csrf().disable()
                .authorizeRequests()
                .antMatchers("/**/*.css", "/**/*.js","/h2-console/**",
"/signup", "/register").permitAll()
                .anyRequest().authenticated()
                .and()
                .formLogin()
                .loginPage("/login")
                .defaultSuccessUrl("/home")
                .permitAll()
                .and()
                .logout()
                .logoutSuccessUrl("/login?logout")
                .permitAll()
                // 作用于 h2-console
                .and().headers().frameOptions().disable();;
    }

    @Override
    protected void configure(AuthenticationManagerBuilder auth) throws Exception {
        //用户信息使用数据库作为数据源
        auth.userDetailsService(userDetailsService).passwordEncoder(passwordEncoder());
    }
```

```
/**
 * 默认返回 BCryptPasswordEncoder 的 PasswordEncoder
 */
@Bean
public PasswordEncoder passwordEncoder() {
    return new BCryptPasswordEncoder();
}
```

其中 UserDetailsService 的实现类 UserDetailServiceImpl.java 如下：

```
@Service
@RequiredArgsConstructor
public class UserDetailServiceImpl implements UserDetailsService {

    private final UserRepository userRepository;

    @Override
    public UserDetails loadUserByUsername(String s) throws UsernameNotFoundException {
        TUser user = userRepository.findByUsername(s);
        return new User(user.getUsername(), user.getPassword(),
Collections.singleton(
    new SimpleGrantedAuthority("USER")));
    }
}
```

10.4.6 云盘主页

云盘主页分为两个部分，分别是上传模块与笔记模块。主页的示例 home.html 如下：

```
01    <!DOCTYPE html>
02    <html lang="en" xmlns="http://www.w3.org/1999/xhtml"
xmlns:th="https://www.thymeleaf.org">
03    <head>
04        <meta charset="utf-8">
05        <meta name="viewport" content="width=device-width, initial-scale=1,
shrink-to-fit=no">
06        <link
href="http://cdn.jsdelivr.net/webjars/bootstrap/4.1.3/css/bootstrap.min.css"
07            th:href="@{/webjars/bootstrap/4.1.3/css/bootstrap.min.css}"
08            rel="stylesheet" media="screen"/>
09        <title>主页</title>
10    </head>
11    <body class="p-3 mb-2 bg-light text-black">
12    <div class="container">
13        <div id="logoutDiv">
14            <form th:action="@{/logout}" method="POST">
15                <button id="logout" type="submit" class="btn btn-secondary float-right">登出</button>
16            </form>
17        </div>
18        <div id="contentDiv" style="clear: right;">
19            <nav style="clear: right;">
```

```
20          <div class="nav nav-tabs" id="nav-tab" role="tablist">
21              <a class="nav-item nav-link active" id="nav-files-tab"
data-toggle="tab" href="#nav-files" role="tab"
22                  aria-controls="nav-files" aria-selected="true">文件</a>
23              <a class="nav-item nav-link" id="nav-notes-tab" data-toggle="tab"
href="#nav-notes" role="tab"
24                  aria-controls="nav-notes" aria-selected="false">笔记</a>
25          </div>
26      </nav>
27      <div class="tab-content" id="nav-tabContent">
……
134     </div>
135 </div>
136
137 <script src="http://cdn.jsdelivr.net/webjars/jquery/3.3.1/jquery.min.js"
138         th:src="@{/webjars/jquery/3.3.1/jquery.min.js}"></script>
139 <!--<script
src="https://cdn.jsdelivr.net/npm/popper.js@1.16.1/dist/umd/popper.min.js"-->
140 <!--
th:src="@{/webjars/popper.js/1.14.3/popper.min.js}"></script>-->
141 <script
src="http://cdn.jsdelivr.net/webjars/bootstrap/4.1.3/bootstrap.min.js"
142         th:src="@{/webjars/bootstrap/4.1.3/js/bootstrap.min.js}"></script>
143 <script th:src="@{/js/popper.min.js}"></script>
144
145
146 <script type="text/javascript">
147     // 打开笔记编辑框
148     function showNoteModal(noteId, noteTitle, noteDescription) {
149         $('#note-id').val(noteId ? noteId : '');
150         $('#note-title').val(noteTitle ? noteTitle : '');
151         $('#note-description').val(noteDescription ? noteDescription : '');
152         $('#noteModal').modal('show');
153     }
154 </script>
155 </body>
156 </html>
```

以上 18~26 行实现了一个导航栏，用于在上传模块与笔记模块间切换。147~153 行实现了一个展示笔记编辑框的 JS 函数。27~134 行为上传模块与笔记模块的实现，内容如下：

```
27      <div class="tab-content" id="nav-tabContent">
28          <div class="tab-pane fade show active" id="nav-files" role="tabpanel"
aria-labelledby="nav-files-tab">
29              <form th:action="@{/files}" enctype="multipart/form-data"
method="POST">
30                  <div class="container">
31                      <div class="row" style="margin: 1em;">
32                          <div class="col-sm-2">
33                              <label for="fileUpload">上传文件:</label>
34                          </div>
35                          <div class="col-sm-6">
36                              <input type="file" class="form-control-file" id="fileUpload"
                                 name="fileUpload">
37                          </div>
```

```html
          <div class="col-sm-4">
            <button type="submit" class="btn btn-dark">上传</button>
          </div>
        </div>
      </div>
    </form>
    <div class="table-responsive">
      <table class="table table-striped" id="fileTable">
        <thead>
        <tr>
          <th style="width: 20%" scope="col"></th>
          <th style="width: 80%" scope="col">文件名</th>
        </tr>
        </thead>
        <tbody>
        <tr th:each="file: ${files}">
          <td>
            <a th:href="@{'/files/'+${file.url}}"
             th:download="${file.filename}"
             class="btn btn-success">下载</a>
            <a th:href="@{/files/delete(id=${file.id})}" class="btn
             btn-danger">删除</a>
          </td>
          <th scope="row" th:text="${file.filename}"></th>
        </tr>
        </tbody>
      </table>
    </div>
  </div>
```

29~43 行实现了一个文件上传表单，选择好文件并单击上传按钮之后，文件将以 multipart/form-data 格式上传至路径为 "/files" 的控制器。44~63 行实现了一个文件展示列表，其中为每一个文件提供了下载与删除的选项。对应操作的文件控制器 FileController.java 如下：

```java
@Controller
@RequiredArgsConstructor
public class FilesController {

    private final FileService fileService;
    private final MinioService minioService;

    @PostMapping("/files")
    public String saveFile(Principal principal, MultipartFile fileUpload) throws Exception {
        //保存文件
        if (fileUpload.isEmpty()) {
            return "redirect:/result?error";
        }
        fileService.addFile(principal, fileUpload);
        return "redirect:/result?success";
    }

    @GetMapping("/files/delete")
    public String deleteNote(@RequestParam("id") long fileId) throws Exception
    {
```

```
        //删除文件
        if (fileService.deleteFile(fileId)) {
            return "redirect:/result?success";
        }
        return "redirect:/result?error";
    }

    @GetMapping(path = "/files/{userName}/{fileName}", produces =
MediaType.APPLICATION_OCTET_STREAM_VALUE)
    public @ResponseBody
    ResponseEntity<Resource> download(Principal principal, @PathVariable String
userName, @PathVariable String fileName) throws IOException,
InvalidResponseException, InvalidKeyException, NoSuchAlgorithmException,
ServerException, ErrorResponseException, XmlParserException,
InsufficientDataException, InternalException {
        //下载文件
        if (!principal.getName().equals(userName)) {
            throw new RuntimeException("权限不足");
        }
        Optional<TFile> fileOptional =
fileService.getFile(principal.getName(), fileName);
        if (fileOptional.isPresent()) {
            //获取文件元数据,构造 getObject 请求
            TFile file = fileOptional.get();
            String objectName = file.getObjectName();
            HttpHeaders headers = new HttpHeaders();
            headers.add(HttpHeaders.CONTENT_DISPOSITION,
                    "attachment; filename=" +
URLEncoder.encode(file.getFilename(),
    StandardCharsets.UTF_8.toString()));
            return ResponseEntity.ok()
                    .headers(headers)
                    .contentType(MediaType.APPLICATION_OCTET_STREAM)
                    .body(new
InputStreamResource(minioService.getObject(objectName)));
        } else {
            throw new RuntimeException("文件不存在");
        }
    }
}
```

与文件操作相关的文件服务类 FileService.java 如下：

```
@Service
@RequiredArgsConstructor
public class FileService {

    private final FileRepository fileRepository;
    private final UserRepository userRepository;
    private final MinioService minioService;

    public void addFile(Principal principal, MultipartFile multipartFile)
throws Exception {
        TUser user = userRepository.findByUsername(principal.getName());
        String objectName = UUID.randomUUID().toString();
```

```java
            //上传文件至MinIO
            ObjectWriteResponse objectWriteResponse =
minioService.uploadObject(objectName, multipartFile.getInputStream());
        TFile file = new TFile()
                .setFileSize(multipartFile.getSize())
                .setObjectName(objectWriteResponse.object())
                .setUrl(String.format("%s/%s", principal.getName(),
multipartFile.getOriginalFilename()))
                .setFilename(multipartFile.getOriginalFilename())
                .setUser(user);
        fileRepository.save(file);
    }

    public boolean deleteFile(long fileId) throws Exception {
        Optional<TFile> fileOptional = fileRepository.findById(fileId);
        if (fileOptional.isPresent()) {
            TFile file = fileOptional.get();
            String objectName = file.getObjectName();
            //删除MinIO上保存的文件
            minioService.removeObject(objectName);
            fileRepository.delete(file);
            return true;
        }
        return false;
    }

    public Optional<TFile> getFile(String userName, String fileName) {
        TUser user = userRepository.findByUsername(userName);
        return fileRepository.findOneByFilenameAndUser(fileName, user);
    }
}
```

笔记相关的页面实现如下：

```html
65         <div class="tab-pane fade" id="nav-notes" role="tabpanel"
aria-labelledby="nav-notes-tab">
66           <button id="newnote" style="margin: 0.25em;" type="button" class="btn
btn-info float-right"
67              onclick="showNoteModal()">
68              + 新增一篇笔记
69           </button>
70
71           <div class="table-responsive">
72             <table class="table table-striped" id="userTable">
73               <thead>
74                 <tr>
75                   <th style="width: 20%" scope="col"></th>
76                   <th style="width: 20%" scope="col">标题</th>
77                   <th style="width: 60%" scope="col">描述</th>
78                 </tr>
79               </thead>
80               <tbody>
81                 <tr th:each="note: ${notes}">
82                   <td>
83                     <button name="edit"
```

```
    th:onclick="javascript:showNoteModal([[${note.id}]],[[${note
    .noteTitle}]],
84                [[${note.noteDescription}]])" type="button"
85                    class="btn btn-success">编辑
86               </button>
87               <a name="delete" th:href="@{/notes/delete(id=${note.id})}"
88                    class="btn btn-danger">删除</a>
89             </td>
90             <th scope="row" th:text="${note.noteTitle}"></th>
91             <td th:text="${note.noteDescription}"></td>
92           </tr>
93         </tbody>
94       </table>
95     </div>
```

以上内容用于展示笔记的列表，并对每一行笔记提供编辑与删除功能的链接。

```
97     <div class="modal fade" id="noteModal" tabindex="-1" role="dialog"
    aria-labelledby="noteModalLabel"
98         aria-hidden="true">
99       <div class="modal-dialog" role="document">
100         <div class="modal-content">
101           <div class="modal-header">
102             <h5 class="modal-title" id="noteModalLabel">笔记</h5>
103             <button type="button" class="close" data-dismiss="modal"
    aria-label="Close">
104               <span aria-hidden="true">&times;</span>
105             </button>
106           </div>
107           <div class="modal-body">
108             <form th:action="@{/notes}" method="POST">
109               <input type="hidden" name="id" id="note-id">
110               <div class="form-group">
111                 <label for="note-title" class="col-form-label">标题</label>
112                 <input type="text" name="noteTitle" class="form-control"
    id="note-title"
113                   maxlength="20" required>
114               </div>
115               <div class="form-group">
116                 <label for="note-description" class="col-form-label">描述
    </label>
117                 <textarea class="form-control" name="noteDescription"
    id="note-description"
118                   rows="5" maxlength="1000" required></textarea>
119               </div>
120               <button id="noteSubmit" type="submit"
    class="d-none"></button>
121             </form>
122           </div>
123           <div class="modal-footer">
124             <button type="button" class="btn btn-secondary"
    data-dismiss="modal">关闭</button>
125             <button id="save-changes" type="button" class="btn btn-primary"
126                 onclick="$('#noteSubmit').click();">保存改动
127             </button>
```

```
128                </div>
129              </div>
130            </div>
131          </div>
132        </div>
133      </div>
134    </div>
```

97~134 行实现了笔记的编辑框,通过编辑其中的内容实现对笔记的新增或更新。编辑框内的内容通过调用函数 showNoteModal 传入。笔记相关的控制器 NoteController.java 如下:

```
@Controller
@RequiredArgsConstructor
public class NotesController {

    private final NoteRepository noteRepository;
    private final UserRepository userRepository;


    @PostMapping("/notes")
    public String createOrUpdateNote(Principal principal, TNote note) {
        TUser user = userRepository.findByUsername(principal.getName());
        note.setUser(user);
        noteRepository.save(note);
        return "redirect:/result?success";
    }

    @GetMapping("/notes/delete")
    public String deleteNote(@RequestParam("id") long noteId) {
        Optional<TNote> noteOptional = noteRepository.findById(noteId);
        if (noteOptional.isPresent()) {
            noteRepository.delete(noteOptional.get());
            return "redirect:/result?success";
        }
        return "redirect:/result?error";
    }
}
```

createOrUpdateNote 方法接收到的 note 对象中如果包含 id,则更新对应的笔记信息,否则将新增一条信息。

10.4.7 页面配置

以上控制器都未提供对页面的无逻辑直接跳转,这项功能将被集中到 Mvc 页面配置中。配置示例 MvcConfig.java 如下:

```
@Configuration
public class MvcConfig implements WebMvcConfigurer {

    public void addViewControllers(ViewControllerRegistry registry) {
        registry.addViewController("/login").setViewName("login");
        registry.addViewController("/result").setViewName("result");
        registry.addViewController("/signup").setViewName("signup");
```

```java
    }

    @Override
    public void addResourceHandlers(ResourceHandlerRegistry registry) {
        registry
                .addResourceHandler("/**/*.css", "/**/*.js")
                .addResourceLocations("classpath:/static/");
    }
}
```

其中结果提示页面 result.html 内容如下：

```html
<!DOCTYPE html>
<html lang="en" xmlns="http://www.w3.org/1999/xhtml"
xmlns:th="https://www.thymeleaf.org">
    <head>
        <meta charset="utf-8">
        <meta name="viewport" content="width=device-width, initial-scale=1, shrink-to-fit=no">
        <link href="http://cdn.jsdelivr.net/webjars/bootstrap/4.1.3/css/bootstrap.min.css"
              th:href="@{/webjars/bootstrap/4.1.3/css/bootstrap.min.css}"
              rel="stylesheet" media="screen"/>
        <title>结果</title>
    </head>
    <body class="p-3 mb-2 bg-light text-black">
        <div class="container justify-content-center w-50 p-3" style="margin-top: 5em;">
            <div th:if="${param.success}" class="alert alert-success fill-parent">
                <h1 class="display-5">成功</h1>
                <span>操作成功，请点 <a th:href="@{/home}">这里</a> 以继续。</span>
            </div>
            <div th:if="${param.error}" class="alert alert-danger fill-parent">
                <h1 class="display-5">失败</h1>
                <span>操作失败，请点 <a th:href="@{/home}">这里</a> 以继续。</span>
            </div>
        </div>
    </body>
</html>
```

全局的异常处理 CloudStorageControllerAdvice.java：

```java
@ControllerAdvice
@Slf4j
public class CloudStorageControllerAdvice {

    @ResponseBody
    @ExceptionHandler(value = MethodArgumentNotValidException.class)
    public ResponseEntity<Result<String>> errorHandler(MethodArgumentNotValidException e) {
        String errorMsg = e.getBindingResult().getAllErrors().get(0).getDefaultMessage();
        log.error("未处理异常" + errorMsg);
        return new ResponseEntity<>(new Result<String>()
                .setMessage("参数错误: " + errorMsg), HttpStatus.BAD_REQUEST);
```

        }
    }

响应实体 Result.java：

```java
@Accessors(chain = true)
@Setter
@Getter
public class Result<T> {
    private int code;
    private T data;
    private String message;

    private final static int SUCCESS = 0;
    private final static int FAIL = -1;

    public static <T> Result<T> ok(T data) {
        return new Result<T>()
                .setCode(SUCCESS)
                .setData(data);
    }

    public static <T> Result<T> failed(String message) {
        return new Result<T>()
                .setCode(FAIL)
                .setMessage(message);
    }
}
```

10.5　测试与验证

测试与验证环节分为集成测试与手工测试两个部分。集成测试关注文件上传以及添加笔记的相关功能。手工测试用于测试完整业务流程，验证程序最终实现是否满足需求。

10.5.1　集成测试

集成测试部分主要测试文件上传以及添加笔记的相关功能。集成测试内容如下 CloudStorageApplicationTests.java：

```java
@SpringBootTest
class CloudStorageApplicationTests {

    @Autowired
    private WebApplicationContext context;
    private MockMvc mvc;

    @BeforeEach
    public void setup() {
```

```java
        //spring security测试相关配置
        mvc = MockMvcBuilders
                .webAppContextSetup(context)
                .apply(springSecurity())
                .build();
    }

    @Test
    @WithMockUser(username = "admin")
    public void testUpload() throws Exception {
        //测试文件上传
        MockMultipartFile file = new MockMultipartFile("fileUpload",
                "filename.txt", "text/plain", "some xml".getBytes());
        mvc.perform(multipart("/files").file(file))
                //期望的结果为转发,并且转发结果是"redirect:/result?success"
                .andExpect(status().is3xxRedirection())
                .andExpect(view().name("redirect:/result?success"));
    }

    @Test
    @WithMockUser(username = "admin")
    public void testAddNote() throws Exception {
        //测试添加笔记
        mvc.perform(post("/notes")
                .contentType(MediaType.APPLICATION_FORM_URLENCODED)
                .param("noteTitle", "title")
                .param("noteDescription", "description"))
                .andExpect(status().is3xxRedirection())
                .andExpect(view().name("redirect:/result?success"));
    }
}
```

10.5.2 手工测试

手工测试的步骤按照一个普通用户访问程序的基本流程进行。访问项目的主页将自动跳转至登录页面,如图 10.5 所示。

图 10.5 登录界面

尝试不输入账号和密码直接登录，页面输入框内将弹出输入提示，如图 10.6 所示。

图 10.6　登录提示

单击"注册"按钮进入注册页面，如图 10.7 所示。

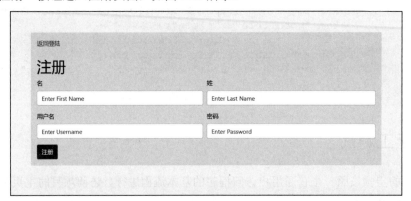

图 10.7　注册页面

尝试不输入内容直接单击"注册"按钮，输入框将弹出输入提示，如图 10.8 所示。

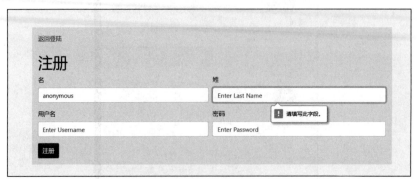

图 10.8　注册提示

输入错误的注册信息，将弹出错误提示，如图 10.9 所示。

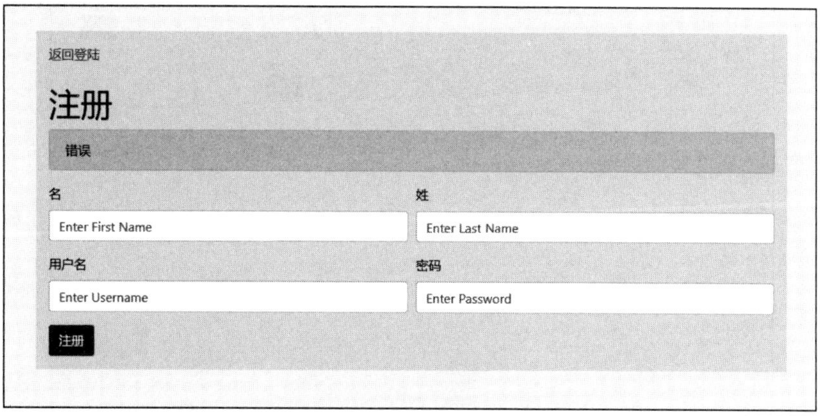

图 10.9　注册错误提示

输入有效的注册信息，单击"注册"按钮后将弹出成功注册以及跳转提示，如图 10.10 所示。

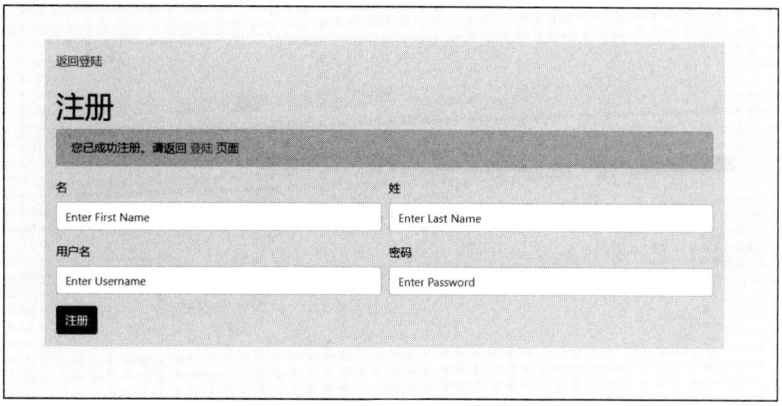

图 10.10　注册错误提示

输入有效的用户账号和密码并登录，如图 10.11 所示。

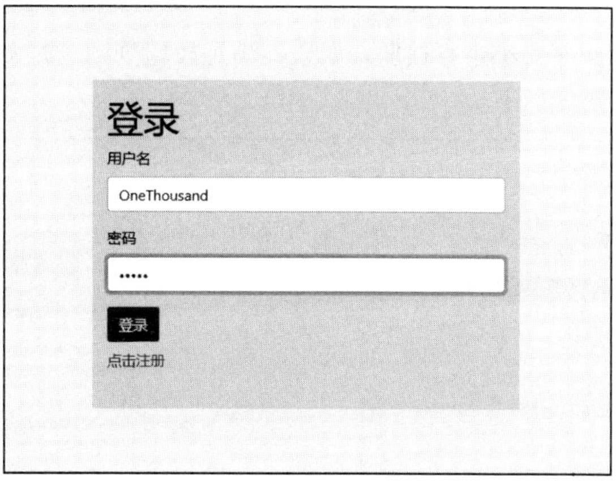

图 10.11　有效的用户账号和密码

进入用户界面后,首先能看到的是文件列表界面,如图 10.12 所示。

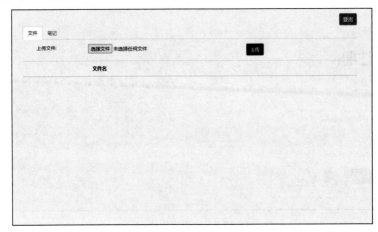

图 10.12　文件列表界面

单击选择文件按钮将弹出文件选择框,以单选的形式选择文件,如图 10.13 所示。

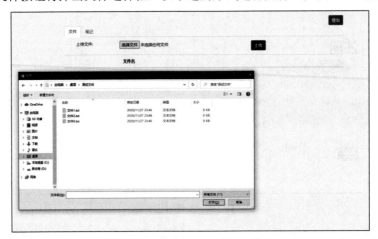

图 10.13　选择文件

文件上传控制器将对文件大小进行校验,如果不满足校验规则(如文件大小为 0),将跳转至失败页面,如图 10.14 所示。

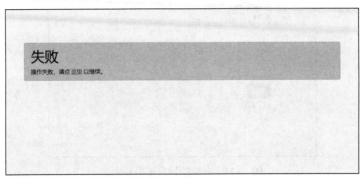

图 10.14　失败界面

如果上传文件满足规则，将跳转至成功页面，如图 10.15 所示。

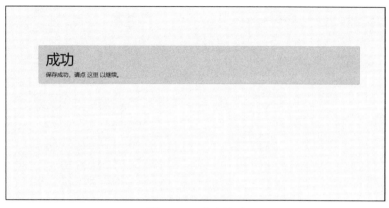

图 10.15　成功界面

上传成功的文件将出现在文件列表中，对每个文件都提供下载以及删除选项，如图 10.16 所示。

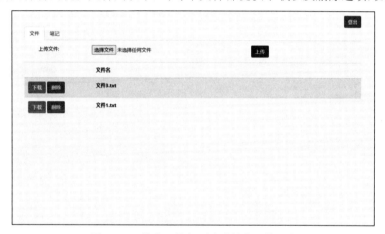

图 10.16　带有下载与删除功能的文件列表

单击"下载"按钮文件将自动保存到本地，如图 10.17 所示。

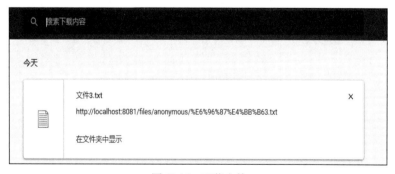

图 10.17　下载文件

单击"删除"按钮文件将从列表中移除，对象存储中的文件也将一并移除，如图 10.18 所示。

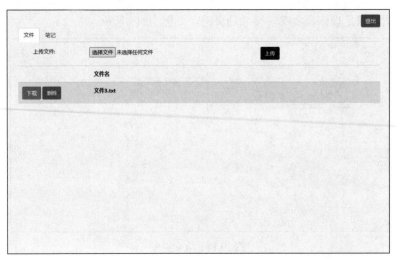

图 10.18 删除文件

单击"笔记"选项卡,页面将切换至笔记页面,如图 10.19 所示。

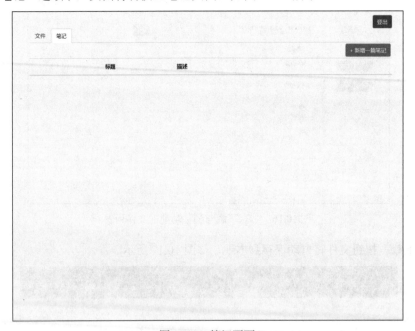

图 10.19 笔记页面

单击"新增一篇笔记"按钮,页面将弹出输入框界面,如图 10.20 所示。

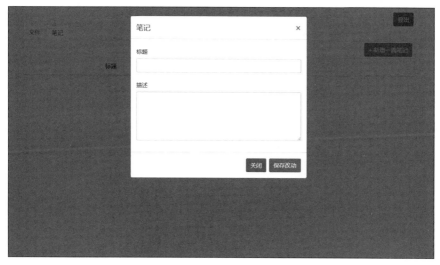

图 10.20 笔记输入框

输入框将验证用户输入,如果不满足输入规则,将弹出输入提示,如图 10.21 所示。

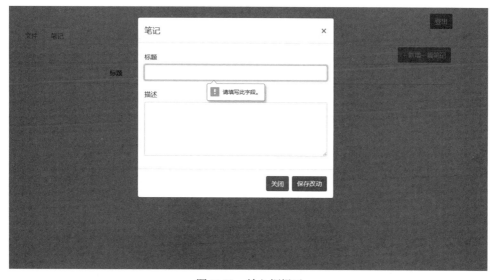

图 10.21 输入框提示

单击"保存改动"按钮之后,笔记将被新增至笔记列表中,如图 10.22 所示。

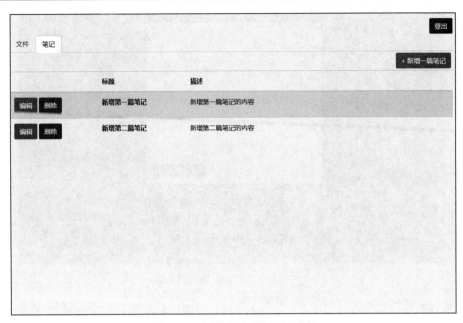

图 10.22　新增成功后的笔记列表

单击"编辑"按钮,然后在输入框内进行编辑并保存,笔记内容将会被更新,如图 10.23 所示。

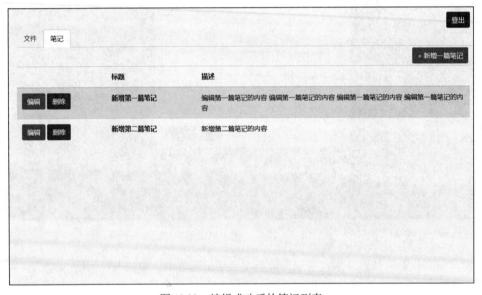

图 10.23　编辑成功后的笔记列表

单击"删除"按钮,笔记内容将从文件列表移除,如图 10.24 所示。

图 10.24　删除成功后的笔记列表

单击"登出"按钮，将回到登录页面，并提示成功登出，如图 10.25 所示。

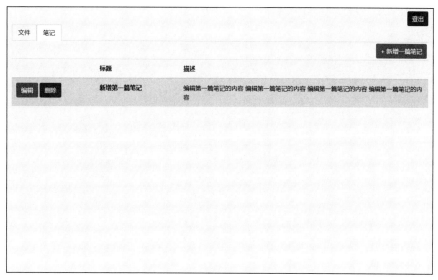

图 10.25　登出系统